Getting into

Engineering Courses

Getting into guides

Getting into Art & Design Courses, 10th edition
Getting into Business & Economics Courses, 12th edition
Getting into Dental School, 10th edition
Getting into Engineering Courses, 4th edition
Getting into Law, 11th edition
Getting into Medical School: 2018 Entry, 22nd edition
Getting into Oxford & Cambridge: 2018 Entry, 20th edition
Getting into Pharmacy and Pharmacology Courses, 1st edition
Getting into Physiotherapy Courses, 8th edition
Getting into Psychology Courses, 11th edition
Getting into Veterinary School, 11th edition
How to Complete Your UCAS Application: 2018 Entry, 29th edition

Getting into

Engineering Courses

James Burnett

4th edition

trotman t

Getting into Engineering Courses

This fourth edition published in 2017 by Trotman Education, an imprint of Crimson Publishing Ltd, 19–21c Charles Street, Bath BA1 1HX

Author: James Burnett

1st–3rd edns: James Burnett

British Library Cataloguing in Publication Data
A catalogue record of this book is available from the British Library.

ISBN: 978 1 91106 732 0

Printed and bound in Malta by Gutenberg Press Ltd

Contents

About the author

James Burnett is the International Director of the MPW Schools and Colleges and advises students on the best routes to gain places at their chosen universities. He has written and edited a number of the Trotman/MPW guides, including *Getting into Art & Design Courses*. He delivers seminars and workshops throughout the world on UK university entrance, the most recent being in Malaysia, Indonesia, Thailand and Singapore. Previously, James was a teacher of A level Physics and A level Photography and, for the most part, managed not to confuse the two. In his spare time he is a photographer and artist, and has exhibited his photographs and multimedia work in the UK and Asia.

Acknowledgements

I am indebted to Lawrence Coates, Professor of Engineering at the University of East Anglia and Admissions Tutor for UEA's Energy Engineering and Integrated Engineering courses, who has generously contributed a significant amount of new material to this edition. He has combined expertise, passion for the discipline and practical advice in a way that will inform and engage the potential engineers who will read this book. Thank you also to Alan Cordner and Christelle Clements-Davis. I am extremely grateful to Holly McIntosh, Mayu Mizuno, Jordan Massiah, Gholamhossein Malekmadani, David Baretto, Aliya Foster, and to Barry Sullivan and Dr Stepan Lucyszyn who contributed to the previous edition of this book. This book contains advice from a variety of professionals, institutions and those who are studying or working within the field of engineering. I have also scattered my opinions, based on over 20 years' experience of helping students with university applications, throughout the book and fully accept responsibility for any errors that appear.

James Burnett
November 2016

Introduction

In the summer of 2016, one of the most popular exhibitions in London was of the work and career of the structural engineer and designer Ove Arup. Ove Arup's company has been involved in many engineering icons, such as the Sydney Opera House and the Pompidou Centre in Paris. It is significant that a national museum that relies on the entrance fees of thousands of visitors in order to help to maintain its collections and staffing should be confident enough in the current interest in engineering to put on the exhibition. However, what is really remarkable is that the exhibition was not in London's world famous Science Museum, but across the road at the Victoria and Albert Museum, a design and crafts museum that won the title 2016 Museum of The Year and whose recent exhibitions have included subjects as diverse as the history of underwear, the clothes of David Bowie, and the psychedelic sixties.

Although engineering touches all aspects of our lives, it has never had as high a profile as, for instance, architecture, medicine or economics among the public. However, in recent years that has changed, as engineering projects ranging in scale from nanotechnology and the electronics that create your smartphone to amazing feats such as sending the Rosetta spacecraft to Comet 67P have captured the public's imagination. Engineers are also involved in environmental and safety issues, as well as helping to shape the world's future energy and raw materials needs.

If you are interested in engineering, you might be able to name some famous engineers – Thomas Edison, Isambard Kingdom Brunel, James Dyson, George Stephenson and Tim Berners-Lee, for example – but most engineers work as part of teams; teams that work on making our lives as fulfilling, safe and manageable as possible without people being aware of what they do. There would be no exciting new buildings or bridges without the structural engineers to work alongside the architects; no iPhones or iPads without the electronic, telecommunication and materials engineers; no websites without software engineers; and Rosetta would not have started its mission without the mechanical, chemical and aerospace engineers who created it and successfully launched it from Earth.

As you are reading this introduction, you are probably thinking about what opportunities exist for engineers. I hope that reading this book, coupled with your research into universities and careers within engineering, will show you that there is a myriad of exciting possibilities waiting for you within this field.

About this book

The aim of this book is to take you through the process of applying to study engineering, from choosing courses and universities through to postgraduate courses and career opportunities.

Chapter 1 focuses on what studying engineering at university entails, including the structure of degree courses, the methods of assessment and the different options for course length and type.

Chapters 2 looks at the desirability of gaining work experience or participating in taster courses and how to go about finding suitable placements.

Chapter 3 covers the process of how to choose the right engineering course for you. There is information on how to start your search and what factors (academic and non-academic) you should use to filter your choices in order to end up with the five courses for the UCAS form.

Chapters 4 and 5 give you the essential information that you need when using the UCAS system to apply, and writing the all-important UCAS personal statement.

Some universities require students to attend an interview and **Chapter 6** discusses how to prepare for them and what to expect.

Chapter 7 contains advice for students who decide to study engineering as a change of direction, mature applicants, students with disabilities and those who are applying from outside the UK.

In **Chapter 8** there is a breakdown of the options available when you get your examination results and what to do if you do not achieve the grades you require.

Financial information, both the cost of studying and what financial support may be available, is covered in **Chapter 9**.

Chapter 10 looks at options for further study and training for graduates, and includes information on career opportunities and Chartered Engineer status.

Chapter 11 lists sources of further information for potential engineers, and at the end there is a glossary of common terms used in this book.

You can use this book as a source of advice and information by reading the chapters or sections that are of relevance to you. However, I recommend that you read the book from the beginning rather than dipping in and out of it, because you will then get a more complete picture.

References to university entrance requirements throughout the book are usually given in terms of A level or AS grades (where appropriate), and the equivalent entry requirements for students studying Cambridge Pre-U, the International Baccalaureate (IB), Scottish Highers and other qualifications can be found by using the UCAS Tariff (see Chapter 3). Universities, on their websites, provide further details of entrance requirements for all of the commonly accepted examinations, and the 'Find a course' facility on the UCAS website (www.ucas.com) will also list these. Regardless of the examination system you are using, the advice on applications given in the book is applicable to all candidates. If you have any questions about your own particular situation, the universities are happy to deal with individual queries and can be contacted via their 'contact us' sections on their websites.

Engineering defined

What is engineering?

What is the difference between science and engineering? There are many definitions, but, essentially, engineering is the practical application of mathematics and science to create machines, processes or structures. Whereas the starting point in science generally involves trying to explain or predict phenomena through the development and verification of theories and models, engineering is the process of physically achieving a goal by applying scientific ideas and theories in a practical way.

What is an engineer?

'*Dreamer. Innovator. Researcher. Problem Solver. Inventor. Creator.*'

(www.whatisengineering.com)

What do engineers do?

Although we tend to classify engineers and engineering courses under different headings, there is a good deal of overlap, and most engineering projects or processes require the input of many different types of engineer. When you look in more detail at the course content of different engineering programmes, you will see that there are many common elements to these. For instance, mechanical engineers will spend time studying electronic and electrical engineering as machines often use electricity as a power source; and civil engineering requires an understanding of the properties of materials, as does structural engineering.

Biomedical engineers

Biomedical engineering (along with biotechnology) looks at the engineering aspects of living things, often for medical purposes. This can range from working with living materials, such as animal tissue, to the development of medical instrumentation (medical scanners and equipment used in surgery), and the design of machines and devices such as heart pacemakers and artificial limbs.

Chemical engineers

Chemical engineering deals with the industrial processes that produce, for example, drugs, food and fuels. Chemical engineers are concerned with not only the chemical properties of the materials they are producing or developing, but also the economic and safety aspects of the projects. Chemical engineering links closely with bioengineering, biomedical engineering and biotechnology.

Civil engineers

Civil engineering deals with the large-scale infrastructure that is an essential part of daily life, such as roads, bridges, dams, water supplies, and office and apartment blocks. (The name civil engineering came about in order to differentiate projects that were there to benefit society in general from military engineering projects.)

Design and product engineers

Design and product engineering deals with the process of creating and developing systems and devices. As well as looking at production processes, such as the design of production lines and factories, design and product engineers have to be aware of safety and cost issues. As an example, the production of a new car involves mechanical and electrical engineering input, computer hardware and software development, the choice of the right materials to ensure that the car will be safe, functional and attractive, and the creation of a production line that will be both efficient and cost-effective.

Electrical and electronic engineers

Electrical and electronic engineers work with electrical and electronic devices. As with mechanical engineering, these range from a microscopic scale (integrated circuits or solid state devices, for example) through to national electricity networks. Many universities offer a joint electrical/electronic engineering course, but it is also possible to specialise in just one of these subjects. There are close links between electronic engineering and computer or information technology (IT) engineering.

Energy engineers

Focusing on the energy industry, energy engineers work on electricity generation and transmission with an emphasis on alternative sources of energy, energy efficiency and environmental issues. Energy engineering is a rapidly growing field, encompassing traditional (fossil fuels, nuclear fission) and alternative (wind turbines, wave and tidal schemes, solar power, biofuels) sources of energy.

Information systems engineers

Information systems engineering is closely linked to electronic engineering. It focuses on computer systems and the transfer of electronic and digital information – mobile phones, the internet and computer operating systems.

Materials engineers

Materials engineering is closely related to structural engineering in that it looks at the properties of materials that are necessary to create structures. However, it also covers the use of materials for other requirements, such as plastics, ceramics, glass and polymers. A mobile phone manufacturer, for example, may work with materials engineers to ensure that the phone is strong enough to withstand daily knocks while at the same time being light and attractive.

Mechanical engineers

Mechanical engineers work in the development and manufacture of machines. This is obviously a very broad description: the word 'machine' covers an enormous range of devices, from medical equipment that is used to perform microsurgery through to aircraft carriers. Mechanical engineering courses include a number of specialist areas, such as aeronautical engineering and automotive engineering. Aeronautical engineers work on all aspects of aircraft and spacecraft, from aerodynamics to the design of engines. Automotive engineers can also work on things that fly, as well as cars and other forms of transport.

Petroleum engineers

Petroleum engineering covers all aspects of the oil, gas and petroleum industries – from exploration and excavation, through refining and purification, to distribution. It covers geological studies, the chemical properties of hydrocarbons, and industrial processes.

Process (systems) engineers

Chemical, mining, petroleum, biotechnology and pharmaceutical engineering can all be described as being branches of process engineering. Process engineers use chemical and biological processes to make raw materials into useful commodities. Process systems engineers tend to work on IT and computer programmes to facilitate these processes.

Structural engineers

Structural engineering deals with the use and suitability of materials that are intended for creating structures such as buildings, bridges, sporting facilities and electricity pylons. The discipline covers the structure and properties of materials on a microscopic level, and the behaviour of structures on a macroscopic scale. Structural engineers work closely with architects and civil engineers. Designing a breathtaking structure based on an architect's sketch can be a very creative and rewarding process, as can handing it over to the civil engineers to build. The career can also be really attractive to those who like using modern computer aided design tools.

There are many other classifications of engineering disciplines and courses (see page 31), and there are subsets of some of the areas mentioned above and courses that combine two or more of these disciplines. So, you will need to spend some time researching possible courses before deciding what you want to do. Advice on this can be found in Chapter 3.

Most engineering projects involve the input of a range of specialisms (see Figures 1 and 2).

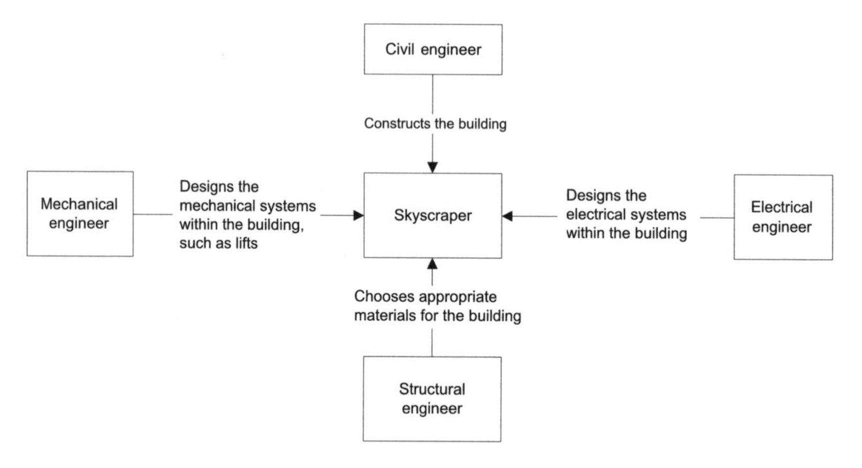

Figure 1: Input into building a skyscraper

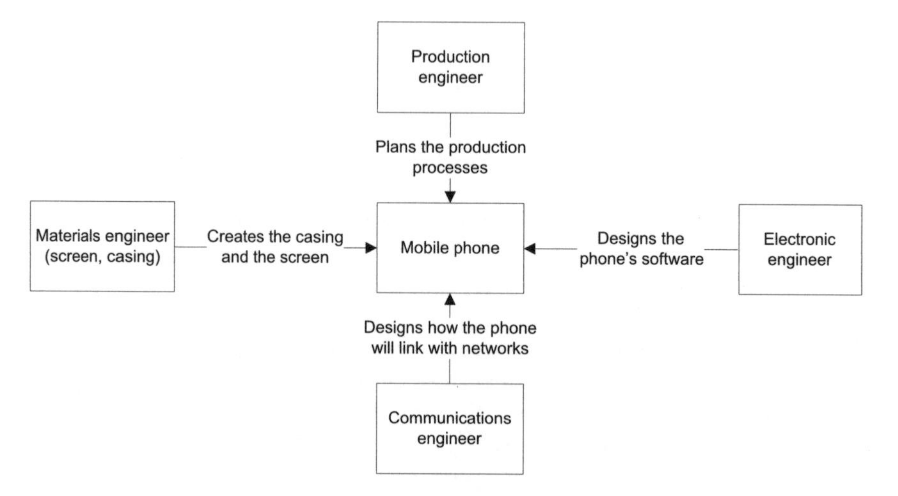

Figure 2: Input into building a mobile phone

Opportunities for engineers

Very few careers provide as many opportunities as engineering while at the same time offering secure employment prospects.

- Engineers can work anywhere in the world.
- The work can be theoretical or practical, and can be carried out within an office or on-site.
- Engineers can work for multinational companies or set up their own business.
- Engineers can work on any scale, from nanotechnology and micro-electronics, through to building the world's biggest structures.

Case studies

Case studies are an interesting and informative way to find out more about engineering. Most university websites will have case studies of graduates, and the professional engineering bodies (see Chapter 11) illustrate their particular disciplines using case studies. The Royal Academy of Engineering's website has a good selection and would make an ideal starting point for those researching a career in engineering: www.raeng.org.uk/education/what-is-engineering/engineering-case-studies.

Case study

Around the world chasing black gold

'After spending time working in the aerospace, automotive and research industries during my studies I had decided they were not for me. Graduating with a master's in Mechanical, Design and Manufacturing Engineering from The Queen's University of Belfast in 1997, I packed my bags and headed to northeast Scotland to commence training as a drilling engineer in the upstream oil and gas industry. My new company was one of the largest oil and gas companies in the world with a reputation for high-quality training, technically strong personnel and a global presence that I hoped would translate into global opportunities for me.

'I spent the next two and a half years in less than sunny Aberdeen developing my wellbore design and operational skills in both the traditional onshore office environment and on offshore oil production platforms and floating drilling rigs. In 2000 I got my first overseas posting to the very sunny (but then war-torn) Luanda, Angola. With this came the opportunity to work on what was then considered to be the new frontier for the industry, ultra-deep water exploration, well designs and drilling operations. The deeper the water, the bigger the design challenges for both the equipment used to drill the wellbores and the equipment installed to construct the wellbores.

'After a couple of years spent living in Luanda I was ready for a fresh technical challenge. This one was to be another industry frontier, High Pressure High Temperature (HPHT) well design and drilling operations. The bags were packed (all two of them and one four-foot wooden rhino) once again and home was set up in the then mystical and beautiful city of Baku, Azerbaijan. The next two and a half years were spent onshore designing and supporting the drilling of some of the world's most prolific gas producing wells. The hydrostatic challenges of the deep water wells had been replaced with the pressure and temperature challenges of these HPHT wells, along with geological challenges such as seafloor instability, active mud volcanoes and tectonically stressed formations. These wells would eventually provide the first gas to Turkey from Azerbaijan via a pipeline that was in itself a world-class engineering feat.

'After seven years designing wellbores and supporting the drilling of them I decided to take the plunge into full-time offshore drilling supervision. I moved back to Scotland and commenced working as a drilling supervisor 75 kilometres west of the Shetland Islands on an oil platform. I was now the person accountable for ensuring the wellbore was drilled and constructed as per the wellbore design and drilling program (previously, I had been the one designing the wellbore and writing the drilling program).

'From 2006 to 2011, I worked once again on HPHT wells, but this time in the Mediterranean Sea, offshore Egypt. I worked offshore as a drilling supervisor to access subsea exploration targets previously considered unreachable, several regional depth records were set and many challenges faced – some we overcame and some we didn't.

'Having notched up seven years' offshore I packed the bags again and moved to Rio de Janeiro, Brazil were I took the position of engineering team leader. I was responsible for all company in-country engineering in support of our deep water exploration. The hydrostatic challenges returned with some new geological challenges thrown in as a result of the primary targets lying below hundreds of metres of deep, highly mobile salt.

'After a fantastic two-and-a-half years in Brazil it was time to return to where it had all started for me overseas, and I moved back to Angola as the Exploration Drilling Superintendent. I was given responsibility for a wonderful toy in the shape of a 6th generation drillship running at US$2million/day to execute a pre-salt exploration wells campaign.

'In 2015 I hit the big 40 years old and decided it was time to take that often discussed, but never planned time out and spend some quality time with my family in our home in Thailand. To be continued …'

Alan Cordner MEng

Working internationally

Few careers provide greater opportunities to work or study overseas, and this is one of the attractions of engineering. Many of the world's major engineering projects are undertaken by multinational companies, and many engineers work freelance, choosing their projects and locations to suit their skills and circumstances. The rapid technological advances of the BRICS nations (Brazil, Russia, India, China and South Africa) have created enormous demands for qualified engineers.

Competition for places

While the ratio of applicants to places for engineering courses is lower than for many other subjects, the competition for places at the higher-ranked universities is intense. Thus many candidates, while being successful at gaining a place on an engineering degree course, do so either through Clearing or at one of their lower-preference universities. You should aim as high as you can (within the boundaries set by your examination results and predictions) in terms of your choice of university, as employers will look not only at what you studied but where you studied. Therefore, do all that you can to ensure that your choice of university will stand you in good stead in the future. As of 2015 entry and beyond, the government has removed the cap on places available, but the top universities are still constrained by space and laboratory/workshop facilities; and although the removal of the cap has given them more flexibility, there are still far more applicants than places at these institutions.

UCAS (www.ucas.com) reports that for the 2016 application cycle around 31,000 students gained places on engineering courses. Of these, approximately 26,000 were men and 5,000 were women. The most popular areas of engineering were mechanical engineering (8,600) and electronic/electrical engineering (4,900).

Application ratios (the number of applications per place) vary from university to university and course to course. Most universities publish these on their websites. For example, the University of Bristol (www.bristol.ac.uk) had 7.6 applicants per engineering place, Southampton (www.soton.ac.uk) had 9 applicants per place for electronic and electrical engineering, UCL (www.ucl.ac.uk) had 5 applicants per place for mechanical engineering and Manchester (www.manchester.ac.uk) had a total of 1,900 applicants for chemical engineering for its 295 places.

As with most things in life, the more planning and preparation you do, the better your chances of success. This is, of course, true for your studies and examinations; but it is also the case that research and a well-planned UCAS application will give you a much better chance of being made offers by your chosen universities. This applies to all aspects of the application: choosing your courses and universities, looking at their entrance requirements, writing a personal statement that will demonstrate your seriousness about, and suitability for, the course, and ensuring that you are prepared for any interviews or entrance tests.

Ignore those who tell you that 'if you are lucky, you will get an offer' – if you pitch your applications at the level that is appropriate for your qualifications (past and predicted) and prepare properly, you can remove most of the uncertainty from your application. The following chapters will tell you how to do this.

1 | Studying engineering

What does studying engineering actually entail? Is it all practical work? How much mathematics is involved? Should I take a joint honours degree? This chapter covers these and many other questions. Once you have read the chapter, my advice is to go to the university websites as they will provide you with more detail as well as case histories and comments from current and past students.

Engineering courses

The structure of an engineering course will vary not only from discipline to discipline, but also from university to university. The common elements to all engineering courses are:

- mathematics (see page 16)
- physical laws (for example Newton's laws or the laws of energy conservation or thermodynamics)
- the physics of materials (physical, electrical and thermal properties)
- environmental, safety and health issues
- cost and other economic issues.

The difference between school and degree courses

'The students who succeed at university are usually distinguished more by their attitude to independent study, their passion for the subject and their willingness to get involved in other aspects of university life than they are by their entry grades. Engineering students work hard and play hard, and they frequently hold senior positions on student society committees or in volunteering groups.

'Staff often have to help students to see that bite-sized answers to questions that might have been appropriate at A level need to become well-argued and researched responses. Just because the question is only one sentence doesn't mean that the answer should be.

'A new development across the sector over the last 10 or 20 years has been the recognition that it is not enough to teach engineering students a multitude of technical skills. There is an increasing amount of discussion around professional responsibility, dealing with ethical dilemmas, learning from past disasters, etc. (see

Chapter 6), which, if taught well, can be just as stimulating as the technical material. The Royal Academy of Engineering has been helping to pioneer these aspects and its website links to many useful publications in this area. So, a student who has studied mathematics, further mathematics and physics may have to adjust slightly as they realise that the broader skills they learned earlier in their school life will need to be dusted off and brought to bear.'

Lawrence Coates, Professor of Engineering, University of East Anglia

Student experience

'At A level there was an average of 20 students in each class, but at university there are 150 students with you in the lecture. If you don't ask questions or don't let the teacher know that you didn't understand something they will just carry on and won't wait for you! It's much easier to lose your concentration and do something else in university and most of the teachers don't really care if you are listening to them or not! We are not chased for coursework or assignments, so you have to be self-disciplined in the way that you study. The lecturers are there to teach you and help you, and if you ask something, they will do everything possible to ensure you understand the concept. In university lectures you will just be taught the basics of the modules and then they will give you resources and you have to study and learn by yourself. There are labs and tutorials to help you figure out your problems, but it is mostly about your own work.'

Gholamhossein Malekmadani, University College London, BEng Mechanical Engineering with Business Finance

Student experience

'During my first year of studying mechanical engineering I had a very full timetable. There were some slight fluctuations with the timetabling each week depending on whether or not we had a practical lab session; the lectures however, remained more consistent. I had very long days for the most part of the week, with some short days that were reserved for lab sessions. I was pleasantly surprised that a lot of the course content was a recap or slight extension of A level material, which made it very easy to settle into the work. The modules were mainly graded on coursework, lab reports and in-class tests, so that at the end of the year there were only two final exams. The assignments were spaced out very well, and we were given two weeks to complete all lab reports, so that even if two pieces of work overlapped, there was still ample time to get the first one out of the way before starting on the next. All lab reports and most assignments were completed as part of a group which helped bring everybody on the course closer together. I greatly benefited from the fortnightly tutorial sessions that took place in classrooms where we would work through exam questions in small groups, with the lecturer giving us fantastic condensed notes and examples.'

Aliya Foster, Kingston University, MEng Mechanical Engineering

Student experience

'I found the difference between A levels and university to be quite stark. At school, most of the learning takes place under supervised conditions and you are made to do lots of repetition and practice to help you learn. This is not the case in university because the teaching is done in lectures, so the reading, the practice and the learning generally take place without any supervision or direct guidance. The process of learning is much more independent at university because you are expected to know how to learn from your A level years and the staff have less time to answer questions and explain things outside lectures. At university there is often no immediate consequence for not completing work, but you must avoid letting yourself fall behind because you will have to do twice as much work to catch-up!'

Jordan Massiah, Magdalene College, University of Cambridge, MEng Engineering

Look at the course content in detail on the university websites to ensure that your interests are covered. Two examples of the typical range of topics covered as part of a four-year (MEng) engineering degree are shown below.

Mechanical Engineering

Year 1

- Mechanics
- Thermodynamics
- Design materials
- Electronics and instrumentation
- Fluid mechanics
- Engineering skills

Year 2

- Engineering systems and control
- Fluids
- Solid materials
- Thermodynamics
- Design
- Manufacturing

Year 3

- Heat transfer
- Structural mechanics
- Materials selection
- Group business design project

Year 4

Range of options including:

- Aircraft propulsion
- Aerospace structures
- Vehicle design
- Robotics
- Energy and environmental design
- Product development

Electrical and Electronic Engineering with Management

Years 1 and 2

- Circuits
- Digital electronics
- Semiconductors
- Analogue electronics
- Thermodynamics
- Software engineering
- Mathematics
- Design and build projects
- Communication systems

Year 3
- Industrial placement

Range of options including:
- Artificial intelligence
- Biomedical engineering
- Control engineering
- Robotics
- Instrumentation
- Optical electronics

Year 4
- Design project
- Range of options (see Year 3)
- Management
- Finance
- Economics

Project work

Project work forms an important part of the later years of a degree course, and is arguably the most exciting part of the course. Imperial College, on its website (www.imperial.ac.uk) says of project work: 'This provides you with the opportunity to demonstrate independence and originality, to plan and organise a large project over a long period, and to put into practice some of the techniques you have learnt throughout the course. It should be the most satisfying and inspiring piece of work in your degree.' The differences between coursework/projects at school and at university are illustrated below.

A level or equivalent	Degree level
• Your teacher helps you because he or she is judged on your results. • Study guides are consulted and marking guidance is utilised. • Marks are maximised by ticking all the right boxes. • Basically your teacher tells you what to do. • Generally there is one right way to do it.	• Your lecturer provides you with a framework within which the coursework is completed. • Some constraints are explained in the handouts. • You interpret them and add value or originality. • Basically your lecturer never tells you what to do. • Generally there are multiple right ways to do it.

Lawrence Coates, Professor of Engineering, University of East Anglia

The importance of mathematics

All engineers use mathematical methods as an integral part of their work. While much of engineering relies on physical processes to develop and produce devices, machines, structures, fuels or chemicals, these are all underpinned by mathematical calculations and models. If you look at the course outlines for engineering degrees you will notice that a significant amount of the first-year content involves mathematics. Most universities will specify an A level (or equivalent) mathematics grade in their entrance requirements, so if you are interested in becoming an engineer you will need to study mathematics. If you want to be involved in production or design but do not want to study mathematics, you could look at alternative courses such as product design.

Very few universities require students to have an A level in further mathematics or the equivalent in order to be considered for engineering courses because some schools don't offer it as an option.

'Mathematics is essential and takes time to learn. A student who is competent in mathematics can probably learn any other engineering subject comparatively quickly. Frequently, when students ask partners from the engineering industry how much of the mathematics that they studied at university they actually use in practice, they are told 'about 10%'. This is a most misleading response. Mathematics trains students and graduates to think logically in everything that they do, including writing technical reports. This is one of the reasons why engineering graduates are sought after by so many different sectors of industry. So, although they may not use Laplace Transforms every day, the fact that they studied them in depth once will have developed their methodical approach to their work that avoids them making mistakes.

'Most universities provide additional mathematics support to engineering students. At the University of East Anglia (UEA) our Student Support Services employs experienced maths tutors as part of the learning support team, and they in turn co-ordinate a group of senior students who also offer help and guidance. So, a student who is prepared to work hard to develop their mathematics skills will usually find there is plenty of help available.'

Lawrence Coates, Professor of Engineering,
University of East Anglia

Methods of assessment and study

Most universities award degrees on the basis of examinations that are sat throughout the course, although the weightings between examinations sat in the different years of the course may vary from university to university. Most engineering courses also involve coursework, dissertations or practical assessments. Details are given on the university websites. Courses will also contain different amounts of laboratory or practical work, work placements with engineering companies, or on-site work experience. This will also be detailed in the course outlines on the university websites and so you can make sure that your chosen course suits your own preferences or requirements before applying. Teaching is normally conducted through lectures (sometimes supplemented by one-to-one tutorials) or laboratory workshops.

BEng vs MEng courses

BEng courses are generally three years in length, and the outcome is a bachelor's degree in the relevant area of engineering. MEng courses are generally four years in length and integrate the bachelor's degree with a master's degree. This is discussed in more detail in Chapter 10. Students who study MEng courses have a more direct route to Chartered Engineer status (see page 111) and will usually command a higher starting salary as well as gaining transferable skills such as teamwork and problem solving, because the master's component involves more group and project work. In most cases, students opting for the three-year BEng course have the opportunity to transfer on to the MEng course after Year 2.

Combined honours courses

There are a number of joint or combined honours degrees available, but there is less flexibility with engineering courses than there might be with arts or humanities subjects. This is simply because engineering degrees tend to lead towards careers in engineering, and so they focus on this.

Many engineers end up running their own engineering company or business, or taking a managerial role within an engineering firm, and so a wide range of degrees that combine engineering with management are available.

As an example, take the Electronic and Electrical Engineering with Management course offered by Imperial College.

The programme for the first two years concentrates on the theoretical and technical aspects of electrical and electronic engineering. In the third and fourth years (there are four years because this is an MEng, not a BEng course, see page 112) there is more flexibility for you to steer the degree towards your own interests, and you would study business courses alongside the engineering courses.

If you intend to apply for a joint or combined honours course, you must ensure that your personal statement addresses both aspects of the course (see page 49).

Other examples of combined honours courses:

- Electronic Engineering and Computer Science – Aston University
- Electronic Engineering and Music – Bangor University
- Energy Engineering with Environmental Management – University of East Anglia
- Civil and Architectural Engineering – University of Bath

Other courses

Foundation degrees

Foundation degrees are two-year full-time (or three-year part-time or distance learning) courses that are provided by some universities. (Do not confuse these with the foundation courses offered to some international students in place of A levels or the equivalent.) They are intended for students who do not have conventional academic backgrounds, for example students who left school after taking their GCSEs and have been working in a relevant field, or mature applicants. Many employers will accept Foundation degrees as an acceptable qualification; and there are many opportunities for students with a Foundation degree to follow this with an extra year of university study to gain a bachelor's degree. You can apply for Foundation degrees through UCAS (www.ucas.com). Subjects available at Foundation degree level include all the major engineering fields.

Higher National Diplomas

Higher national diplomas (HNDs) are usually two-year courses, often equivalent to, or taught simultaneously with, the first two years of a bachelor's degree. Students who are successful on the HND course can study for a third year to gain a bachelor's degree. Entrance requirements are normally less stringent than for a degree. For example, Coventry University asks for BCC (104 UCAS Tariff points) from three A levels (including A level Mathematics and/or Physics) for entrance onto some degree-level courses, but only 80 points from two A levels, for the equivalent HND course.

Scottish degree courses

Undergraduate degree courses at Scottish universities leading to bachelor's degree qualifications are four years in duration, although it is sometimes possible to enter in year two. This is sometimes called 'Advanced Entry' and the advantage of this is that it reduces the course length to three years for a BEng course or four years for the MEng course. The Scottish universities that offer this option will specify the entry requirements for the second-year (level 2) entry on their websites. The grade or score requirements are generally higher than for the first-year (level 1) entry. For example, Strathclyde University asks for BBB at A level for first-year entry onto the Electronic and Electrical Engineering course, and ABB for entry into the second year.

The course structure of the four-year degree allows students to study a broader range of subjects in the first year, compared with what is on offer in the three-year degrees common in England, Wales and Northern Ireland. The MEng qualification (see page 17) normally takes five years in Scottish universities.

2 | Getting work experience

Work experience that is related in some way to engineering is an invaluable way of demonstrating to the universities that you are committed to the course. It also shows that you have researched how engineers translate their academic studies into the practical skills that are required in the real world. Work experience will also show you whether you are suitable for a career in engineering, and what qualities are needed in a successful engineer. One of the things that an admissions tutor will look for is how serious you are about your chosen course. By writing about your work experience and how what you saw relates to what you enjoy studying or to things that you have read about, you can show the selectors that you have thought about engineering as a whole and as a potential career, rather than as an academic discipline only.

Many surveys have highlighted the importance that engineering employers attach to internships and work experience. And this is also true of your university application. As more and more students chase a fixed number of places, preference is given to those candidates who can demonstrate that they have made an effort to find out what working within the field of engineering will be like.

Engineering affects every aspect of our lives, and engineers make things that need to be commercially viable, and so gaining work experience that is in some way related to engineering should not be difficult. If you are lucky enough to have some connection with an engineering firm then use this as a starting point. But there are many other part-time jobs and work placements that will allow you to investigate or experience aspects of engineering, such as:

- any company that physically makes something that is then sold commercially – light bulbs, toys, jam jars, window frames, car accessories
- farm work
- shops that sell or repair computers or mobile phones
- a local architecture practice
- a local builder.

What will you gain from work experience?

- You will have lots to write about in your UCAS personal statement, and you will be able to demonstrate your research into engineering.
- It will help you to decide which area of engineering is most suitable for you. Do you want to work outside or in an office? On practical problems or on the theoretical side of the subject? In a small business or a large multinational company?
- The contacts that you make during your work experience may be helpful in your future career.
- You will find out whether you would actually *enjoy* working as an engineer. Be aware of the quote from the American inventor and engineer Thomas Edison: 'I never did a day's work in my life. It was all fun.' If you don't think being an engineer would be fun then it is probably not the right career for you.

Looking for work experience

Where to start

You may be fortunate in that your school will arrange this for you as part of a work-experience scheme. If not, then you will need to look for placements yourself.

So, how do you get work experience?

- You could approach local companies or use any contacts that your family may have.
- The institutes of engineering have contacts with engineering companies so try to go through them (see contact details in Chapter 11). Some will also offer summer courses or 'taster' programmes. The Year in Industry programme provides help and advice for students aiming to use their gap year to gain work experience (www.etrust.org.uk/the-year-in-industry/about-yini).
- There are many schemes operated by universities and the engineering institutes aimed at attracting students into engineering, and these often involve work placements.

If you know someone, or of someone, in a local engineering firm, try asking to go in for one or two weeks' work experience or work-shadowing during the holidays. Remember that even a single day of work-shadowing is better than no evidence of experience within the workplace, and something even tangentially related to engineering is better than no work experience at all. Helping organise the files for a local car mechanic would give you some access to the practicalities of engineering problems, and would be a good stimulus for reading more about automotive engineering, for example.

Summer and 'taster' courses

Another good way to learn more about engineering is to use part of your summer holiday on an organised programme that gives you a taste of what a career in engineering entails. A Headstart (www.etrust.org.uk) course is one of many options available and is highly recommended by university admissions tutors. The UCAS website's taster course search page is a good starting point: www.ucas.com/events/exploring-university/learn-about-uni-taster-course.

How to apply for work experience

You need to prepare a curriculum vitae (CV), sometimes called a résumé. This should be short and to the point, outlining your education, experience, achievements and contact details. It should be word-processed on plain paper (but can be accompanied by a handwritten letter), and it should not be longer than two sides.

Your CV should include:

- full name, address, telephone number and email address
- education – places, qualifications and grades (start with the most recent)
- skills (e.g. computer skills, software packages you are familiar with, languages spoken, whether you hold a driving licence)
- work experience (full time or part time, with names and addresses of the companies or businesses and a brief description of your responsibilities)
- positions of responsibility
- hobbies
- names and contact details of two or three people who can act as your referees.

Points to remember:

- highlight any experiences or achievements that are relevant to engineering
- highlight any skills gained (teamwork, communication, responsibilities)
- ensure that the layout makes it easy to read

- check all spelling and grammar carefully
- make sure there are no gaps in the CV (periods of time that are unaccounted for).

A sample CV

Lay out your CV clearly and logically, avoiding gaps, and including any exams you are studying for as well as those taken. Below is an example.

Jonathan Luke

Address: 1 Cameron Road, Richmond, London TW9 1MB

Telephone: 0123 456 7890
Email: jl@whizzmail.co.uk

Education
2011–18: Richmond High School
2018: A levels to be taken: Physics, Chemistry, Mathematics
2016: GCSEs: English (A), Mathematics (A), Geography (A), German (A), Biology (B), Chemistry (B), History (C), Physics (C)

Work experience
2016–17 (Saturdays)
Sales Assistant in Heggie's Department Store.
Responsible for operating the checkouts during busy times; dealing with customer enquiries and complaints; checking till receipts against takings.
The job requires good communication skills, the ability to deal sympathetically with complaints, and accuracy in dealing with the financial aspects of the post.

Skills
Modern languages: good written and spoken German.
IT: competent in MS Word, PowerPoint and Excel; good keyboard skills.

Positions of responsibility
Captain of school volleyball team; treasurer for the school film society.

Interests
Volleyball, swimming, reading, film, travel and music.

References
Available on request.

The covering letter

The covering letter (often handwritten, unless the company asks for a word-processed letter) should add more detail to the CV and also explain why the post you are applying for is suitable for you. You should find out the name of the person who will read it, and refer to them by name. If you use the person's name, sign off with 'Yours sincerely'. If for some reason you need to address it to an unknown person, then use 'Dear Sir' or 'Dear Madam' and end it with 'Yours faithfully'. If you are asked to email your application, send the letter and CV as attachments and give a brief overview (name, why you are looking for work experience, brief details of current studies) in the body of the email.

A sample covering letter is shown below.

Mr Yorkston
Head of Personnel
Hopkins Engineering
Fence Road
Edinburgh

23 January 2017

Dear Mr Yorkston

I am interested in applying for a part-time post at Hopkins Engineering, to gain some work experience in preparation for my application for a degree course at Fife University. I am particularly interested in your company because of its specialism in GPS systems, which is something I have been researching as part of my International Baccalaureate studies, for my extended essay. I am hoping to study electronic engineering at university next year.

In addition to my qualifications shown on the attached CV, I have been working part-time in my local charity shop and I am often left in charge and so I have learned the importance of taking responsibility. At school, I have formed a science club and invite speakers to come to talk to us once a month. Incidentally, Dr Tunstall, who is one of your research engineers, came to talk to us last month about developments in GPS systems, and we found her talk extremely informative.

I hope that you will be able to consider me for a post. I can be contacted by telephone or email at your earliest convenience.

Yours sincerely

Kenny Anderson

Work experience interviews

If you are lucky, you may then be called for an interview. Most of the advice given in Chapter 6, on university interviews, is also relevant to job interviews. Other points to remember are listed below.

- Do some research on the company in advance. Know about what they produce and how long they have been operating.
- Ensure that you can explain how your particular skills and qualities would be useful to them.
- Be clear about when you can start, how long you can work for, and whether there are particular periods when you cannot work, for example on the day your examination results are released.
- Wear smart clothes, a business suit if possible. Ensure that you have polished your shoes and don't wear too much jewellery.
- Introduce yourself, address the main interviewer by name, and offer your hand for a handshake at the start and again at the end.

Student experience

'I found work experience incredibly useful as it helped me to confirm my interest in mechanical engineering and gave me an inside look into the day-to-day life of an engineer. Although I knew about what engineers did, I had never been able to experience it first-hand, and work experience gave me a unique insight into how projects come together and are completed. I found the knowledge that I gained from this very helpful when writing my personal statement and applying to universities, as well as useful for me in clarifying my future plans.

'I did placements at two different engineering firms and an innovation centre, all of which I organised by researching the companies and then reaching out via email to them. I enjoyed my work experience and would recommend that anyone wanting to study engineering undertakes some.'

Holly McIntosh, UCAS applicant

How to use your work experience effectively

Undertaking work experience is a means to an end – to show the university selectors that you are a serious candidate. You will be expected to write about what you discovered about being an engineer in your personal statement, and to discuss it in more detail at your interview. It is useful to keep a diary of your experiences while you are there. Record what you did, saw and heard, and in particular conversations you had with engineers – not just technical things, but about their training and what they see as the main challenges and rewards in working in an engineering field. It is also a good idea to then do some research or reading on things that you encountered during your work placement.

3 | Choosing your course

Choosing universities and engineering courses can be a bewildering experience because you will be confronted by an enormous number of options. This chapter covers the steps that you should take to narrow your search and to find the most appropriate courses for you, including what to look for in the course content of an engineering course, academic requirements, types of university and location. It also covers what the course or university can offer you in terms of your future career and what you will gain from it on a personal level.

What to consider

Academic considerations

You are allowed five choices on the UCAS application. The basic factors to consider when choosing your degree course are:

- the engineering course you are looking for
- where you want to study
- the kind of university you are aiming for
- your academic ability.

Non-academic considerations to bear in mind include:

- accommodation
- location
- sporting, musical or other extracurricular facilities
- costs.

You need to think seriously about your choice of universities, as the decisions you will take now may determine your future career options. At this stage, you may already have an idea of which universities you want to consider, based on the advice of friends and/or family, but you need to be as open-minded as possible. Make a list of between 10 and 20 universities in which you are interested; it is then important to reduce this to a much shorter list. Not only will the university be where you begin the next step in your education, it will also be your home for three or four years, so think carefully about the location, environment and accommodation options as well as the suitability of the course(s) on offer.

Here are some things to research and consider for each university.

- Get hold of the prospectuses and any departmental brochures for more details. Remember that university publications are there to attract applicants as well as to provide information, and may be selective about the information they provide, so read all of it bearing this in mind.
- Visit the websites of the universities you are considering. This is the best place to look for the most current information about a university. Another useful element to university websites is information on past and present students from a range of disciplines who give their views on student life at the institution. Some university websites have email links to current students who can answer any questions that you may have.
- Find out when the open days are and go to them if you possibly can. You will have a chance to look at the facilities, talk to current students and find out more about the course.
- Discuss engineering and studying it as a subject with people you know who work as, or with, engineers; ask for their views on the reputations of different universities and courses. This may bring up some highly rated engineering institutions you may not have thought of.
- Investigate the grade requirements and be realistic about the grades you are expecting – your teachers at school or college will be able to advise you on this.
- Check that the course allows you to choose the particular options in which you are interested. If you are considering, for example, mechanical engineering but have a particular interest in aeronautics or automotive engineering, make sure that these options are available. You will not always know what each option actually covers by its title, so read the department's own prospectus carefully and address any unanswered questions by contacting the admissions tutors directly – contact details are usually on the departmental website.
- Think about whether you would like your course to include a placement with an engineering company. This might be for a few weeks, a term or even a year. If this is something that interests you, find out who organises the placement – you or the university – and whether there is any funding to cover, for example, travelling costs or subsistence.
- Consider whether you want to spend some time abroad. Many engineering courses offer the option of a year abroad, studying at a partner university. If so, think about practical details such as language or visa requirements, and costs.
- Investigate how much practical or laboratory work is included in the course, and what are the practical engineering facilities. Find out whether the facilities include state-of-the-art machinery or testing equipment.

- Look at what IT facilities are offered. If you do not have your own laptop, will the university have facilities for you to manage without one of your own? Access to computing facilities can be very important when you are working on a dissertation. Find out whether there is WiFi access in the study areas and/or accommodation.
- Ask about the reading lists, and whether the books are available in the library. Are you required to buy your own textbooks? If so, are second-hand copies available?
- Look up the specialties of the engineering course staff. Are they experts in the field of engineering in which you are particularly interested? Use the internet to find out what their experience is and what they have published, as this will give you a better indication of a department's strengths.

Student experience

'I was looking forward to living and studying in London, firstly because of the high quality of the universities. I liked London because, besides student life and studying hard, you can have fun. Then I looked at the course content of the mechanical engineering courses to see what suited me. My other selection criteria were the university rankings and then their ranking for my course. Then I asked for advice from some professors in the UK and in my home country of Iran, and finally talked with my friends who studied in these universities and got further advice from them.'

Gholamhossein Malekmadani, University College London, BEng Mechanical Engineering with Business Finance

League tables

Newspapers often feature university rankings or league tables but there is no official ranking of universities or university courses in the UK, and so these tables are created using criteria selected by the newspapers themselves. There is a significant amount of variation between these tables, because each table will score the universities in a different way.

However, as long as you approach these rankings with caution, these tables can be a useful aid to the selection process, particularly if you look at how the rankings are assessed rather than simply looking at a university's position in the tables. There will be some criteria that you might regard as being important to you – graduate job prospects, for example – while you might not be so interested in the student to teacher ratios.

A good starting point is the *Guardian* newspaper's university rankings. The general rankings can be found at: www.theguardian.com/education.

And there is a link to the subject-specific rankings which feature a number of engineering disciplines. Of particular interest might be the scores based on job prospects, and the student satisfaction with the course and with the teaching.

You might be interested in how the UK's universities are regarded on the global stage, particularly if you are an international student or you are planning on working abroad at some stage in your career. There are a number of world university rankings which you might find useful. But bear in mind that, like the UK university rankings, they are not in any way 'official' rankings and are based on criteria that the organisations compiling them see as being important.

One example is the world university rankings compiled by *Times Higher Education* (www.timeshighereducation.com).

There are many other league tables and rankings, for example the one produced by *The Sunday Times* newspaper for UK universities, or by QS, which specialises in business education, for world rankings. No two rankings will produce the same results and so you need to use them as part of the process in making your decisions, not as the sole reason for a choice of university.

Another useful reference point is the National Student Survey, which offers a student perspective on individual degree courses at any given university. The questionnaire is completed by graduates across the UK; the results are published annually and can be viewed online at www.hefce.ac.uk/lt/nss/results.

After completing your research, you should be able to narrow down your original list to the five choices for your UCAS form. Once you have done this, discuss the list with your teachers to see whether they think it includes sensible choices. They may ask you to think again about some of the choices.

Common areas of concern for a school's UCAS adviser are:

- your current subjects don't fit the requirements of the courses you are applying for
- that the grade requirements are too high (or too low) for the applicant's likely academic achievements
- that there is too much variation within the choice of courses to enable the applicant to write a coherent and focused personal statement (see Chapter 5).

Choosing the right course

Most universities offer a wide range of engineering courses, and it is important for you to investigate these thoroughly before making your choices. A particular university might list the following on the UCAS website:

- Aeronautical engineering
- Aeronautical engineering with a year abroad
- Biomedical engineering
- Biomedical engineering with a year abroad
- Chemical engineering
- Chemical engineering with a year abroad
- Chemical with nuclear engineering
- Civil engineering
- Civil engineering with a year abroad
- Electrical and electronic engineering
- Electrical and electronic engineering with a year abroad
- Electrical and electronic engineering with management
- Information systems engineering
- Materials science and engineering
- Mechanical engineering
- Mechanical engineering with a year abroad
- Structural engineering

A broad outline of general engineering courses can be found on page 11, but see the university websites and prospectuses for more details on specialisms. You will need to spend some time going through these. Some universities offer integrated or general engineering courses that allow students to choose their specialisation later in the course. If you are truly undecided about which particular area of engineering is the most suitable for you then this might be a good option. However, even if you apply for an integrated or general engineering course, it would be a good idea to have an idea of which specialisations interest you most as this will probably be asked at the interview.

When considering possible courses, read through all of the course content. Do not choose a course just because of its title. Courses with the same name at different universities can vary immensely in their content, and within the courses themselves the likelihood is that you will have a range of options to choose from once you start your course. This is also important if you are interviewed (see Chapter 6) because you may be asked to justify your choice. Being able to discuss the course structure in detail will be an important factor in convincing the interviewer that you are a serious applicant.

Similar-sounding courses also do not always have the same entrance requirements (examination results and preferred A level subjects). Examination results are specified as either grade requirements (for example, AAB) or Tariff points (for example, 136 – see the section on the new UCAS Tariff below). Unless you are applying post-results (as a mature applicant or during your gap year), your referee will be asked to predict the grades that you are expected to achieve in your examinations. You should find out in advance what he or she is going to predict, because this will determine your choice of universities and courses. For example, if you apply for five university courses that require AAB but your A level predictions are BBB, there will be a high chance of being rejected by all of your choices. You will then have to try to find alternatives through the UCAS Extra scheme, or through Clearing (see Chapter 8). Similarly, if you are predicted to achieve A*AA, you are probably aiming too low if all of the courses you are applying for require CCC at A level.

As a rough guide, if you are predicted, say, ABB, it would be risky to choose a course that requires AAB. It would be safer to choose three or four that require ABB, and one or two that require BBB. This not only means that you have a good chance of getting a number of offers, but it also gives you options if you do not quite meet the grade requirements (see Chapter 8).

Sandwich courses

Sandwich courses, either BEng or MEng, offer students the chance to spend some time gaining industrial experience through work placements organised or suggested by the university. Typically, this might involve taking a year away from the university, although some universities will also offer a variety of shorter placements. Students who choose sandwich courses do so because:

- they want to gain experience of working to enhance their future job prospects
- they want to learn more about what working as an engineer entails
- they want to make contacts in the engineering world to help them to find jobs in the future.

Overseas study

Studying abroad could also be a factor that affects your degree selection. It is possible to study engineering in many countries as part of a degree based at a British university. Not all of these courses send you off for a full year, though: there are schemes that last for only one term or semester. You do not need to be a linguist either, as it is always possible to study overseas in an English-speaking location such as North America, South Africa, Australia or Malaysia. As an example, if you study Electrical and Mechanical Engineering with International Study at the University of Strathclyde you will have the opportunity to study at a partner university in Year 4. The university's partner universities include institutions in Canada, USA, China, Singapore, New Zealand and Australia.

The availability of student exchanges has increased through programmes such as Erasmus+, which encourage universities to provide international opportunities where practical – particularly in Europe. The popularity of overseas study has encouraged some universities to develop special exchange relationships with universities further afield.

Many universities will provide contact details or testimonials from students who have studied abroad as part of their degrees. For example: www.strath.ac.uk/studywithus/studyabroad/goingabroad/international exchange/studenttestimonials.

Academic and career-related factors

A level reform

A levels in England are in a transition period, with new specifications being phased in for first teaching between September 2015 and September 2017 to replace the 'legacy' modular specifications. 'Legacy' specifications refer to old-style A level qualifications composed of AS and A2 units, where AS and A2 scores had equal weighting towards the overall A level mark. Under the reforms, the new, 'reformed' A levels are linear rather than modular, and the AS and A level have been decoupled, which means that the new AS is now a separate qualification and no longer counts towards the A level. The changes are summarised below:

'Legacy' specifications (modular examinations)	'Reformed' specifications (new)
AS units contributed 50% of the total A level mark, and were usually sat at the end of Year 12	AS can be sat as a stand-alone exam but does not contribute to the A level
Old-style A2 units were sat at the end of Year 13. Unit marks were added to AS marks to give overall grade	A level exams sat at end of Year 13
All schools offered AS (although some schools only allowed the AS units to be sat alongside the old-style A2 units at the end of Year 13)	AS as a stand-alone qualification offered by some schools, but not all

Some subjects moved to the reformed specifications in September 2015, for the first A level examinations in June 2017, while a second group of subjects was introduced for first teaching in September 2016 and will have their first examinations in June 2018. Some subjects (including mathematics) are still following the 'legacy' (modular) specification, with the reformed A level coming into effect for first teaching in September 2017. For more information on what will be available for examination in which year of the course up until Summer 2019, visit www.gov.uk/government/publications/get-the-facts-gcse-and-a-level-reform/get-the-facts-as-and-a-level-reform.

Note that the A level reforms are different in each country of the UK: www.ucas.com/sites/default/files/ucas-guide-to-qualification-reform.pdf. The changes discussed above apply to A levels for students sitting the qualifications through the English examination boards, but there are also separate educational reforms taking place in Wales and Northern Ireland (as listed on the UCAS website).

The UCAS Tariff

While some universities make offers based on grades at A level, Scottish Highers or IB scores, others make offers using the UCAS Tariff. The UCAS Tariff is a points-based system that compares different education systems or achievements. For example, ABB at A level is equivalent to 128 Tariff points, roughly equivalent to an IB Diploma score of 27. There is a new UCAS Tariff in place for September 2017 entry and beyond. If you are familiar with the system operating in previous years, you will be aware that an A* grade at A level, for example, was equivalent to 140 points – it is now 56 points, with an A grade equal to 48 points and 16 points for an E grade. AS scores are now worth 40% (or as close as possible) of the full A level UCAS Tariff points at each grade. H7 in IB is valued at 56 points, and a top grade in a Scottish Higher is 33 points. Full details can be found on the UCAS website at: www.ucas.com/advisers/guides-and-resources/tariff-2017.

Academic ability

It is important to be honest about what you think you will achieve in your A levels or equivalent because, for most, this is the deciding factor for selection. The best way to get a strong sense of your predicted results is to speak to your teachers.

Remember, be realistic; you may think that you can do much better than your teachers' predictions or than your AS (if you sat them as part of the legacy modular A level system or as a standalone qualification under the reformed specifications) results indicate, but the predicted grades and previous results will go on your UCAS application form, and so it is important to ensure that the courses you apply for are consistent with your likely results. If you do much better than expected, you can always 'upgrade' using the UCAS Adjustment scheme (see page 102). Check also whether the universities are likely to make a grade or Tariff points offer. If the grade requirement is, for example, AAB, don't assume that a combination of grades that gives you the same Tariff points (136, achieved by gaining A*BB) would satisfy the university's requirements.

For more details about UCAS and filling in your application, see *How to Complete Your UCAS Application: 2018 Entry* (Trotman Education).

Educational facilities

Take a look at the facilities the university has to offer. Here is a checklist of what to look out for:

- access to lecturers if you need help
- computer facilities
- course materials
- laboratory/workshop provision
- lecture theatres
- library facilities
- multimedia facilities
- study facilities.

Quality of teaching

The Higher Education Funding Council for England, the Higher Education Funding Council for Wales, the Scottish Funding Council and the Department for Education and Learning of Northern Ireland assess the level of teaching across the UK. Their findings are publicly available – see www.hefce.ac.uk, www.hefcw.ac.uk, www.sfc.ac.uk and www.delni.gov.uk. League tables (see page 29) normally incorporate these into their rankings. From September 2017, the government is introducing the Teaching Excellence Framework (TEF), which will assess universities and colleges on the quality of their teaching and is intended to help students with their university choices.

Type of institution

There are three types of institution from which you can obtain a degree:

- 'old' universities
- 'new' universities
- higher education colleges.

The 'old' universities

These are traditionally seen as the most ancient universities (such as Oxford, Cambridge, Edinburgh, University College London, Imperial College and Durham), along with those established in the early and mid-20th century (such as Bristol, Leeds, Liverpool and Reading), plus those established in the 1960s (such as Sussex, Bath and Stirling); usually with higher admission requirements, and with a strong emphasis on research.

The 'new' universities

Pre-1992 these were polytechnics, institutes or colleges, for example Kingston, Central Lancashire and Westminster. These tended to focus more on vocational courses with less emphasis (or none at all) on research. There are a number of excellent engineering degree courses at new universities that are very well regarded and highly competitive to get into, and they are often more flexible in terms of sandwich courses, work placements or links with industry.

Colleges of higher education

These are specialist institutions that have links with universities. The university awards the degree and can deliver part of the course, along with the institution. Many colleges also offer pathway programmes, such as Access or foundation courses for students who do not have the necessary academic qualifications for direct entry to a university.

Non-academic considerations

The starting point for your research ought to be the type of course you want to follow. Once you have done this you should then look at the non-academic facets of the universities that you are considering. These are irrelevant to the course, but instead concern what the university can offer you on a personal, social and cultural level.

Finances

Finance is an important factor to consider, as you will need to juggle a lot of outgoings when you go to university. You will need to take into account:

- accommodation costs
- availability of part-time work in the area to earn some extra money
- living costs, such as food
- travel from your accommodation to the university during term time
- travel from your home to and from the university for holidays
- proximity to your home, family and friends – will it cost you a lot of money to visit friends or to go home?

Accommodation

The university is likely to be your home for three or four years, so think carefully about the location, environment and accommodation options. Accommodation can vary wildly between institutions so you will need to think about where you would feel most comfortable. Do you want to live in halls of residence with other students, or independently (or with friends) in rented houses or flats? Do you want to be near your lectures or are you happy to live further away from the university?

Universities all have accommodation offices that provide help or information for students who do not want to, or are unable to, live in university-provided accommodation. Most universities offer arranged accommodation for first-year students in halls of residence, which may either be owned by the university or be shared with other institutions, but often students are expected to find their own accommodation in subsequent years.

Entertainment

You will be spending the next few years in a new place so you will need to have a look at the entertainment facilities it has to offer both within and outside the university. Are your particular interests or hobbies catered for? If you're a keen sports enthusiast, have a look at the facilities on offer and the sorts of clubs and teams you can join.

Site and size

- Campus university outside a town or city?
- Campus within a town or city?
- University buildings at various locations within a town or city?
- Large or small?

While some students have a clear picture of where they want to study, others are fairly geographically mobile, preferring instead to concentrate on choosing the right degree course and see where they end up. But university life is not going to be solely about academic study. It is truly a growing experience – educationally, socially and culturally – and so you need to do your research and think carefully to ensure that you choose the best university and course for you.

What can the university offer you?

Universities need students. Even the most popular and oversubscribed universities are constantly looking to attract the best students. League tables and advice from teachers, employers or friends can help you in choosing the most suitable courses, but if you plan your application carefully you will have some choices to make. In this chapter we have looked at a range of selection criteria: location, accommodation, the course structure, facilities and other things. You might also want to investigate whether scholarships are available (see Chapter 9) and also the feedback from students who have previously studied at the university. The *Guardian* newspaper's university rankings incorporate scores based on student satisfaction with the course and with the teaching. 'Which? University' (www.university.which.co.uk) also gives statistics on dropout rates, ratios of male:female students and other information that you might find useful.

4 | Completing your UCAS application

This chapter is designed to help you complete your UCAS application. Further advice on filling in your application is given in *How to Complete Your UCAS Application*, which is updated every year.

The UCAS application

The UCAS application is completed using Apply, the online application system on the UCAS website. There are five sections to complete.

1. **Personal information:** name, contact details, home address and nationality (for fee purposes – see Chapter 9).
2. **Your choices of university and course:** you have up to five choices of university or courses. This is entered by using codes for the university and the course. You need to take great care in ensuring that this is accurate. You will also be able to specify the year of entry (are you planning on a gap year?) and whether you want to live at home. Some universities, such as Oxford, Cambridge and Durham, operate a college system so you also have to choose the college at which you wish to be based.
3. **Education:** past examination results, details of your previous and current schools and colleges, and future examinations. This section requires particular care, and you will need to discuss this with your school or college to ensure that, for example, AS unit results are correctly entered if you sat exams as part of the 'legacy' modular system through the English examination boards (see Chapter 3).
4. **Employment:** have you been in full-time employment?
5. **Personal statement:** see Chapter 5.

When you have completed the application, you save it and mark it as complete. You then send it to your referee via the UCAS website. He or she will then complete the reference and, if applicable, add predicted grades or scores. If you are applying as a private candidate, you will give the contact details of a referee and the university will contact him/her directly (see Chapter 7).

University engineering departments are looking for motivated, well-qualified individuals, and so they are keen to provide as much encouragement and practical advice as possible. If you look hard enough, you will find lots of information from the universities that will help you with your application.

All universities provide detailed information on their websites about their selection criteria, including advice on writing the personal statement and an indication of how they assess each candidate.

Suggested timescale

Use the timescale below to help you plan your application.

Year 12

In year 12, during your first year of A levels or IB, you should start to think seriously about what type of engineering course you want to follow. Talk to as many people as possible – your teachers, family and careers advisers. Don't just focus on what area of engineering interests you most, but also on whether you want a three-year BEng course or a four-year MEng course, whether you want a campus university or one in the centre of a city, close to home or in another part of the country.

Your next step should be to find out when the universities that interest you have open days and arrange to visit them. After this, you ought to be in a position to make up a shortlist of courses and universities, ready for your application. You can start to order prospectuses from the universities, or download the PDF versions from the university websites. Remember to make a note of the grade requirements for your chosen universities.

In August, refine your choices in the light of any AS results or predicted examination results.

Year 13

In September of year 13, you can submit your UCAS application. The deadline for applications is 15 January, but I recommend that you apply as early as possible. If you are applying to Oxford or Cambridge universities, your application has to be submitted by 15 October. Some universities will want to interview you, and this can happen from November onwards. If you are required to sit entrance tests, you may be asked to do this in the first week of November.

You should start to receive decisions from the universities from January onwards, although some universities might get back to you earlier than this. You can keep up to date with the status of your applications using the online UCAS Track facility.

If you are unlucky enough to receive rejections from all of your choices, or you decide to withdraw from your choices, you can then use the UCAS Extra scheme from February to apply to other universities, or to add choices if you did not use all of your five choices in the initial application. Extra allows you to add one new choice at a time. If you are successful in gaining an offer from your first Extra choice and you accept it, you are committed to accepting the place; if you decline the offer or are rejected then you can approach another institution through Extra. According to UCAS, around 260 students gained places on engineering courses through the Extra system in the 2016 application cycle.

Once you have received responses from all five universities (or offers through UCAS Extra), you will be given a deadline by UCAS by which time you have to choose one university as your firm acceptance, and a second insurance offer, normally one that requires lower grades. Once you have accepted your first choice university, you will be sent information about accommodation and fees and other practical information directly from the universities.

You sit your examinations in the summer, and receive your results in July or August:

- A level results are published in the third week of August
- Scottish Higher results are released in the first week of August
- IB results come out in the first week of July.

When the exam results are published, UCAS will get in touch and tell you whether your chosen universities have confirmed your conditional offers. Do not be too disappointed if you have not got into your chosen institution; just get in touch with your school/college or careers office and wait until Clearing begins in early July, when all remaining places are filled. You will be sent instructions on Clearing automatically, but it is up to you to get hold of the published lists of available places and to contact the universities directly. UCAS reports that around 3,500 students gained engineering places through Clearing in the 2016 application cycle.

If you have done better than expected, you can use the Adjustment system to look for universities that require higher grades. According to UCAS, 50 students used Adjustment to gain engineering places in the 2016 application cycle.

Entrance examinations

As competition for places is so fierce, some universities ask applicants to sit entrance examinations as part of the application process.

University of Cambridge

There will be written assessments for students applying for 2017 entry onwards. If you are applying for engineering courses at Cambridge you will be asked to sit the Engineering Admissions Assessment (ENGAA) in the November of the year you are applying. You will normally sit the test at your school or college and they will need to register you by 18.00 (UK time) on 15 October. The test has two sections (written and multiple choice) and covers physics and mathematics. Information on the test can be found on the University of Cambridge website: www.undergraduate.study.cam.ac.uk/applying/admissions-assessments/pre-interview.

University of Oxford

Engineering applicants for the University of Oxford have to sit the Physics Aptitude Test (PAT). The test is sat at your school or college in the first week of November. Your school or college will need to register you for the test by 18.00 (UK time) on 15 October.

The paper is two hours long and is divided into two sections: Section A, Mathematics for Physics; and Section B, Physics.

Section B comprises short written answers. A typical question might involve being given the electric current recorded when a box containing electrical components with three terminals is connected in different ways to a power supply. You would have to deduce what the components are and the values of their electrical resistance from the data.

Sample tests and mark schemes can be found at: www2.physics.ox.ac.uk/study-here/undergraduates/applications/physics-aptitude-test-pat.

Other universities

Kingston University, Liverpool John Moores University, Birmingham Metropolitan College and some other universities may ask some candidates to sit an aptitude or admissions test (in the course requirements information on the UCAS website, this is listed as 'IOT', which means 'institution's own test'). The tests are only normally required for mature students or students with unusual qualifications and you will be given information about whether you need to sit the test, and its format, when you apply. The tests are used as one piece of evidence in assessing candidates, alongside many other criteria such as grades achieved, predicted grades, the personal statement and references. The UCAS website provides a list of courses which may require additional admissions tests:www.ucas.com/how-it-all-works/explore-your-options/entry-requirements/admissions-tests.

Taking a gap year: deferred entry

Many students take a gap year between their final school or college examinations and the start of their university course. Universities are nearly always happy with this as students who take a year away from studies are often more motivated and mature when they start their degree studies. But in terms of the application, a gap year will only enhance your chances of getting a place if you use the year productively.

There are two application routes for students taking a gap year.

- You can apply for deferred entry; that is, you apply in the final year of the A level course for entry a year later. So, if you are sitting A levels in June 2017 you would apply for deferred entry in September/October 2018, not 2017.
- Alternatively, you can apply at the start of the gap year, once your A level results are known.

There are advantages to both routes, depending on your plans and A level (or equivalent) grades.

Advantages of deferred entry

- Once you have satisfied your offer, you will know where you will be studying in a year's time, and so you can make firm plans about what you will do during the year.
- You can plan to be overseas, either travelling or on work or voluntary placements, without worrying about having to return for university interviews.
- You have a second chance to apply to universities during the gap year if your initial application is unsuccessful.

Advantages of applying during the gap year

- More time to decide which field of engineering really interests you.
- You will already know your results so you can focus your application on those universities for which you have already achieved the necessary grades or scores.
- If your predicted grades are not high, but you feel confident of doing better than your school or college expects, you can avoid the possibility of being rejected on the basis of your predictions rather than actual ability.

'Applicants are led to believe that universities have differing views on the value of a gap year. This is not likely. A student who is taking a gap year to put off starting university and looks set to waste their time will always be frowned upon. But a student who has clear plans of almost any sort will gain much maturity from a gap year. Within a body of Freshers it is always possible to spot the students who have had a gap year. They are more mature and engage in interesting conversations beyond the obvious 'What did you get in your A levels?'

'A typical gap year involves working for four to six months to earn money to fund travelling. This can be a good use of time. Some applicants engage in activities for charity which can be equally valuable. Anything that exposes you to new experiences will broaden your mind.

'A common concern with taking a gap year is that the student will forget their mathematics, which is such an important part of engineering, particularly in the first year. Even though it should be like riding a bike, there is no doubt that a year completely away from mathematics does mean that gap year students arrive a bit rusty and often do badly in the self-diagnostic tests that are commonly used at entry. But within a couple of weeks they are back up to speed with no long-term issues. So, this is no reason to avoid a gap year.

'Occasionally there are issues to be aware of associated with accommodation guarantees, etc. So, it is worth keeping in touch with the admissions tutor during the gap year. A significant minority of students complete their gap year and then decide to change degree course. Some universities probably use this as a reason for not encouraging gap years, but on the contrary this is a really good thing. If a student has had time to think about career options and has made a positive decision to change (see Chapter 7) then this is far better than arriving at a degree and realising half way through the first year (or even later) that it may not be the right one.

'Keeping in touch with the admissions tutor can help both sides. You may learn some useful information that makes the transition to university easier. The admissions tutor is frequently one of your lecturers and will usually welcome the opportunity to get to know some of the students in advance of their arrival. Occasionally they can put you in touch with relevant student societies so that you are already in contact before the mad rush of Freshers' week. A good example is Engineers without Borders that has branches at many universities and will welcome your attendance at some of their meetings before you arrive.

'Considerable value can be added by keeping a regular log of what you do during your gap year. When completing job applications five years later it can be very helpful to be able to reflect on what you learned from the year; trying to do it solely from memory leads to a less rich account.'

Lawrence Coates, Professor of Engineering, University of East Anglia

Student experience

'I initially wanted to go to university without taking a gap year, but during the weeks before my A level exams I decided against it and went for a deferred entry, as I hadn't got into a London university and wanted to give myself another opportunity. Taking a year out is significant, but there were a lot of things I wanted to do that I could not do while at university, mainly spending time in Colombia (where I was born) and starting a business.

'My first three months away were spent travelling within Colombia and spending time with family. After this period I settled in the capital city of Bogota, and established a routine which included a lot of meditation, fitness and salsa dancing. The remainder of the day I spent building and organising my business, which I had gotten a grasp of while travelling. I did travel a little bit, too, though sparingly.'

D.S. Barreto Romero, University College London, MEng Mechanical Engineering with Business Finance

If you are planning on taking a gap year, you must plan it carefully so that you will gain something from it. This could be life or work experience, maturity, a chance to extend your studies into new areas, independence or money to fund your university studies. A year spent resting and playing computer games, however attractive that may be after all your hard work at school, is not going to convince the universities that you will be a stronger candidate as a result.

Here is an excerpt from a personal statement about taking a gap year:

> *'I am taking a gap year in order to gain more maturity and experience.'*

Such a statement is not going to convince the admissions tutors that you have made constructive plans for your gap year, nor is it likely to help you develop or bring new skills and ideas onto their course. A better statement might be:

'During my gap year I am going to take part in a voluntary project building small-scale dams with Starfish, a voluntary organisation based in Thailand, in a mountainous area, to help control the water that causes soil erosion and crop damage. We will then work in a small village building toilet facilities for the old people who live there. I am looking forward to being practically involved with these engineering projects and to learn more about problem solving in real situations. To raise money for this trip, I will be working in my local hospital in the maintenance department.'

This is much more impressive because the candidate has linked what she will do in her gap year to her future degree course (civil engineering), and it is clear that she has thought carefully about what she will do during the year.

A piece of advice: phrases such as 'I have arranged to ...' are much more convincing than 'I hope to ...' when discussing your gap year plans.

Gap year plans

It is always a good idea to check with your chosen universities that a gap year is acceptable to them before committing yourself. There is likely to be information on their websites addressing this. If there is none, you can email the admissions staff to ask them. This is particularly important if you are taking a gap year for reasons other than wanting to take a year between school studies and the degree course to gain experience in engineering.

Some of the other reasons for taking a gap year are:

- to work towards extra qualifications because you need to strengthen your application, or because you wish to change direction (for example, if you have studied mathematics up to GCSE only, you will be doing an evening class in A level Mathematics alongside your other projects)
- you started another course (such as a degree course in another subject; see Chapter 7) and then realised that it is not right for you, so you have withdrawn from it
- you have been working, and you now want to return to studying
- you may have had an illness or other issues that required you to take a break from studying.

Replies from the universities

After your application has been assessed by the university, you will receive a response. You can also follow the progress of your application using the online Track facility on the UCAS website. You will receive one of three possible responses from each university:

- conditional offer
- unconditional offer
- rejection.

If you receive a conditional offer, you will be told what you need to achieve in your A levels. This could be in grade terms, for example AAB (and the university might specify a particular grade in a particular subject – AAB, with an A in mathematics), or in UCAS Tariff points (136 points from three A levels – see Chapter 3). Unconditional offers can be given to students who have already sat their A levels, such as gap year students applying post-results. Rejection means that you have been unsuccessful in your application to that university.

Once you have received responses from all five universities, you will need to make your choice of the university offer you wish to accept. This is called your firm choice. You can also choose an insurance offer, effectively a second choice with a lower grade requirement. UCAS will give you a deadline of about a month from the date that you received your fifth response to make this decision.

If you receive five rejections, or if you reject all of the offers you have received, then you can enter the UCAS Extra scheme, through which you can make additional choices. See page 41 for more on Extra.

5 | The personal statement

Arguably, the most important part of your UCAS application is the personal statement. It is also the one part of the form where you have complete freedom to decide how you wish to demonstrate your suitability for the course to the selectors. You have 4,000 characters (47 lines) to convince your five chosen universities that:

- you have good reasons for studying engineering
- you have researched your future career thoroughly
- you have appropriate personal and academic qualities to become a successful engineer
- you will be able to contribute something to the department and to the university.

Who will read your personal statement?

Before you can write a personal statement you have to think carefully about your choice of courses. This is because each admissions tutor will read the personal statement with his or her own course in mind, and he or she will expect what you write to be consistent with the course. For this reason, make sure that the five courses you choose are closely linked in terms of course content and outcome. If you are applying for a civil engineering course, the selector will expect to read about your interest in civil engineering, books related to the subject that you have read, and relevant work experience. Similarly, an admissions tutor for electronic engineering will expect the personal statement to address this subject. Clearly, you cannot convince both that you are serious about their courses in one personal statement.

'Your personal statement is your chance to tell me what makes you special. I can already see your academic profile (past and predicted), so this is your opportunity to tell me how you're going to contribute to the Cardiff academic and social community. Is there something that makes you unique and you think we might value? Tell me about it and how you think this is relevant to studying engineering. Lots of students talk about how well rounded they are, and in the eyes of

> many admissions tutors, this is important. But once in a while I would like to read about an applicant who is truly special and unique. Have you done well in the face of adversity? Have you spent time in another culture? A leader? An athlete? How have these experiences made you better, and how will your skills help you to succeed and stand out amongst a group of other high-achieving peers? If you think it is relevant, tell me about it. Don't have anything like this to write about? That's OK too – maybe instead you can write about what challenges you expect to face in the transition to university and how you will overcome them.'
>
> Barry Sullivan, Head of Admissions, Cardiff School of Engineering, Cardiff University

Joint honours courses

Personal statements for joint honours courses are usually read by selectors from both of the departments to which you are applying. If you want to apply for a joint honours engineering and management course, someone from the engineering faculty will expect to read about engineering while their colleagues from the management department will want to read about management. This is fine if you apply to five similar courses, but if you apply for, say, three engineering with management and two single honours engineering courses you will find it difficult to satisfy the selectors from the two different types of course. Discuss your course choices with your careers adviser to assess how best to present these choices in your personal statement.

Applying for different courses at the same university

You can apply for more than one course at a particular university; you do not have to choose five different universities. But applying for two courses at the same university does not necessarily increase your chances of studying there. Take the case of a student who is desperate to study at a particular university, perhaps because she has friends there or she likes the city. She decides to apply for both the civil engineering and biomedical engineering courses. The admissions tutor for civil engineering will look at her application for the civil engineering course, and the admissions tutor for biomedical engineering will look at the application for this course.

Our applicant's main interest is civil engineering, so her personal statement emphasises this, but it also devotes one paragraph to her interest in biomedical issues. The civil engineering admissions tutor reading the personal statement will judge it on how it addresses this course, so he or she might not be fully convinced that the student is serious because the personal statement will not focus enough on the reasons for the choice of civil engineering, and what the candidate has done to investigate it (reading, work experience, etc.). Similarly, biomedical engineering is a very specialist area, and the admissions tutor for this course will expect to read a personal statement that focuses on this, and he or she is not going to be very interested in reading about bridges and roads.

So, by trying to give herself a better chance of getting to this university, the applicant is actually reducing her chances. There are some instances where it is possible to apply for two separate courses at the same university, if they are very similar, but it would be advisable to discuss this with the university admissions department before doing so. So, when writing the personal statement, try to imagine how it will come across to each of the departments to which you are applying. Do not try to write something too general in order to allow yourself the luxury of applying to a wider range of courses.

The structure of the personal statement

What is a perfect personal statement? Of course, there is no such thing. The key to writing a personal statement is to think about the word 'personal' – it is about you and so it has to reflect your strengths, achievements, qualities, research and ambitions. Having said that, there are some elements that are important in creating a successful personal statement:

- reasons for your choice of course
- what you have done to investigate the course and the profession
- what unique qualities and achievements you have
- other things that may affect your application, such as gap year plans.

Reasons for your choice of course

These could include:

- what first got you thinking about engineering, for example watching the news about a new engineering project, an article in a newspaper about new developments in the mobile phone industry, or personal experience such as work experience – this could date back many years, for example how you liked taking things apart and reassembling them when you were very young

- things you have studied at school, for example a topic in physics or chemistry that particularly interested you
- how your particular interests and qualities make engineering a suitable career.

What you have done to investigate engineering

This could include:

- books, articles, magazines or websites that you have read
- work experience (see Chapter 2)
- talks or lectures
- school visits to museums or exhibitions
- the relevance to engineering of things you have studied at school.

Your achievements

These could include:

- academic achievements; for example, Maths Challenge or the Mathematical Olympiad
- extracurricular activities and accomplishments
- responsibilities; for example, sports captain, head of house, chairman of a club or society
- hobbies.

Your personal qualities

These could include:

- teamwork; for example, sports, school orchestra, Management Challenge, Duke of Edinburgh expeditions
- communication skills; for example, voluntary work, school prefect, school council
- leadership; for example, being captain of the school rugby team
- problem solving; for example, competitions, experimental projects, EPQs.

Other information relevant to the application

This could include:

- gap year plans
- if you are an international student, why you want to study in the UK
- what qualities or experiences you could offer to the university or the course.

Blogs and online lectures

Many universities upload lectures so that they are available to the public (including potential students). If you are unable to go to talks and presentations because of school commitments or because you do not live near a university that offers these, then downloading online lectures or following blogs is a good way to learn more about engineering.

Two good examples (Imperial College and the University of Bath) are given below (www.imperial.ac.uk/be-inspired/social-and-multimedia/lectures-online and https://soundcloud.com/uniofbath/sets/public-lecture).

'Preparation for writing the personal statement

1. **Evidence of study of what the subject involves.** This can be anything from indicating some relevant books or websites to having gained work experience with a local company. Good websites, such as the Royal Academy of Engineering, have a lot of useful and reliable explanations about what engineering involves. Increasingly, applicants have been attracted to engineering by watching various programmes on the Discovery Channel, such as 'Extreme Engineering'; this is fine.
2. **Courses and summer schools.** A student who has made the effort to attend a summer school or Headstart course has demonstrated that they have tried to find out what a degree involves. This isn't necessarily the same as what a career involves, but is a good start.
3. **Evidence of an understanding of the breadth of the subject.** Most engineering disciplines are associated in the popular imagination with specific elements, for example mechanical engineering with Formula 1, or civil engineering with structural design. Any evidence of understanding that mechanical engineers also study thermodynamics, efficient engines, sustainable design, etc., or that civil engineers are particularly concerned with water treatment, flow and management, soil mechanics and geotechnics and involve people management, will demonstrate some depth to the background research.
4. **Passion.** It might be argued that an applicant doesn't know enough about the subject to yet be passionate about it. This might have been true 20 years ago, but the internet and YouTube have put an end to that. Any applicant interested in engineering can find a wealth of informative resources that could excite their interest.

5. **Understanding the value of transferable skills.** Sometimes applicants approach the exercise in too narrow a fashion. Usually this manifests itself as a focus on the technical side of engineering and demonstrates a lack of awareness of the broad transferable skills that are needed and developed in a degree. Accordingly, almost any work experience can offer valuable insights into the applicant's approach to team-working and problem-solving. It will not be enough to mention in some vague way that you like working in a team; it would be far better to explain how a specific team-based activity has helped you to understand your strengths and weaknesses.
6. **The value of work experience.** Often applicants are reluctant to mention work experience unless they think it is directly relevant. I have known applicants not mention a Saturday job because it is in a supermarket. However, any job demonstrates experience of the world of work and offers opportunities for the curious student to find out how the business works, even if they are just shelf-stacking.
7. **Studying A levels or equivalent requires commitment.** So, any applicant who is managing to do well at this level and is still maintaining committed contact with organisations such as Scouts, local sports teams or musical productions, for example, is demonstrating the necessary ability to work hard and play hard that is required at university.
8. **Think very carefully about what to write about leisure activities.** You have to remain honest, but if all you do is play games and socialise, now might be a good time to start broadening your activities. Some applicants make the effort to go to local branch meetings of professional engineering societies, others subscribe to industry magazines. All such things demonstrate some commitment to the career. Remember, you are only at university for three or four years; your engineering career is for life. You need to expose yourself to the industry in some way to find out what the job involves.'

Lawrence Coates, Professor of Engineering,
University of East Anglia

Advice from a student

'When writing a personal statement for your UCAS application, my advice would be to first write everything you can think of that is remotely relevant, including every reason you have for wanting to study your chosen subject and all the extra-curricular activities you've done. From there, you can begin to redraft and remove the things you think are less useful to your application. Bear in mind that the majority of your personal statement should be focused on academic issues and only a paragraph should be given to personal hobbies, such as sport and music. When talking about non-academic activities, you should try to explain the skills that this activity has helped you develop. For example, if you play a musical instrument you can explain that it has helped you gain personal discipline – you should explain to the reader how your activities have made you into a better university candidate. Ensure your personal statement is true and that you've read everything you said you had. You should have notes on the books you've read which you can use to refresh yourself on their contents before your interview so that when you are asked about it you have something intelligent to say.

'When preparing for an interview you should always do a mock interview with someone you don't know well to best simulate the environment in which you will find yourself. Also, to prepare for an interview, you should write out model answers to some of the stock interview questions, such as 'Why do you want to study this subject?', 'Why do you want to study at this university?' and 'Why are you suited to study this subject?' Remember, you should not rush your UCAS application, so start preparing as early as you can to maximise your chances.'

Jordan Massiah, Magdalene College, University of Cambridge, MEng Engineering

Analysing a personal statement

I first became interested in engineering during my A level Physics classes [1]. As a child I had always enjoyed taking things apart, although I was not always able to put them back together again.

To further investigate engineering, I spent a week at a local architecture practice, looking at how they worked with a structural engineer to ensure that their designs were practical. I also visited a construction project as part of my school's work experience programme, and I was fascinated to see how things I had read about and studied were used in real life [2]. I have also read as much as I can about engineering, including *A Short History of Engineering Materials* by John Cameron [3]. I have enjoyed going to public lectures on engineering, and talking to engineers, who have given me a much better idea about a career as an engineer [4].

Alongside physics, I am studying mathematics and history of art. Mathematics is an important tool for engineers [5], as well as teaching me to think in a logical way. History of art is an analytical subject, and it puts architecture within a social and political context. It also involves looking at the use of materials, and how architecture and sculpture were able to develop as new materials were introduced [6].

I enjoy sport and music. I am captain of my school football team and so have had to develop leadership and communication skills as well as physical fitness. I play the guitar in a band, and the cello in the school orchestra, which help me with my manual dexterity and teamwork. Outside of school, I enjoy cooking and cycling. I am a member of my local cycling club and compete most weekends.

I believe that my combination of A levels and my research into engineering as a career make engineering an ideal choice for me.

Points raised by this personal statement

An admissions tutor who read this sample personal statement made the following points.

General

While it is clear that the candidate has done some research, there is very little detail in the statement – it is very general – and so I do not really get a clear picture of the depth of knowledge the candidate has about engineering, or about his/her particular areas of interest. It is also, to be honest, a little on the bland side and also a bit frustrating – a lot of sentences which should lead to something that will interest me end without giving me any information.

Specific points (the numbers refer to the relevant passages in the statement)

1. It would have been nice if he/she could have given an example – perhaps it was electricity, or the behaviour of materials, or some problems involving forces?
2. This should have been the most interesting part of the statement.

The candidate could have told me about the links between his/her studies and the work experience. This will tell me that the student has gained something from the work experience, and that he/she is thinking about engineering rather than just doing work experience to look good on the application form.

3. I am always encouraged when students read around the subject, but what I would like to know is, again, some detail. How do the ideas in the book link to A level study and the real world?
4. This would have been an ideal opportunity for the candidate to show me that he/she has been really thinking hard about his/her future career and about whether he/she has the right qualities to be a successful engineer.
5. Give an example.
6. Not many applicants for engineering study history of art, so this immediately makes him/her stand out. And I don't know much about history of art, so an example here would be interesting for me to learn about – I'm sure it would make me want to learn more by meeting the candidate.

A revised (and much better) version of the personal statement, based on the above advice, is now given.

Revised sample personal statement

As a child I had always enjoyed taking things apart, although I was not always able to put them back together again. I first became seriously interested in engineering during my A level Physics classes when we looked at the properties of solid materials, and I began to understand why, for example, the development of reinforced concrete revolutionised the construction industry.

To further investigate engineering, I spent a week at a local architecture practice, looking at how they worked with a structural engineer to ensure that their designs were practical. I also visited a construction project as part of my school's work experience programme, and I was fascinated to see how things I had read about and studied were used in real life. The engineers explained that the concrete girders used to hang the curtain walls of the office block had to be strengthened along their top surfaces because in a cantilever, the tensile forces are at the top and concrete is weaker in tension than it is in compression. I have also read as much as I can about engineering, including *A Short History of Engineering Materials* by John Cameron, and I was fascinated about how the use of cast iron in early 20th-century America enabled architects and engineers to build the prototypes to today's skyscrapers. I have enjoyed going to public lectures on engineering, and talking to engineers, who have given me a much better idea about a career as an engineer. In particular, I began to understand that an engineer needs to be able to analyse information quickly and to be able to solve problems. My aim is to study structural engineering and then to work alongside architects in the creation of exciting new buildings.

Alongside physics, I am studying mathematics and history of art. Mathematics is an important tool for engineers because the starting point of any engineering project is an analysis of its feasibility. The use of integration to find a centre of mass, for example, can help with the design of an asymmetric building. History of art is an analytical subject, and puts architecture within a social and political context. It also involves looking at the use of materials, and how architecture and sculpture were able to develop as new materials were introduced. The transition from the small, squat Romanesque churches to tall and graceful gothic cathedrals in Europe was due to the invention of the flying buttresses, which, in turn, were only effective when tensile forces were reduced by the addition of heavy statues or decorative stone elements.

I enjoy sport and music. I am captain of my school football team and so have had to develop leadership and communication skills as well as physical fitness. I play the guitar in a band, and the cello in the school orchestra, which help me with my manual dexterity and teamwork. Outside of school, I enjoy cooking and cycling. I am a member of my local cycling club and compete most weekends.

I believe that my combination of A levels and my research into engineering as a career make engineering an ideal choice for me.

Adding the extra information requested by this admissions tutor would add detail, make it more interesting for him to read (so he is more likely to want to meet the student), demonstrate that the student is interested enough in the subject to be thinking about links between his studies and what he has experienced, and bring it up to the required length.

Linking your interests and experiences

In your personal statement, try to avoid creating what amounts to a list of things you have read, studied or experienced. It is better to make connections between these in order to demonstrate that you have thought carefully about what is required to be a successful engineer:

- how your A level studies are related to things you will study at degree level
- how something you observed during work experience stimulated further reading
- how skills that you gained from extracurricular activities, such as communication or leadership, are useful for potential engineers.

You could link:

- an article you read about increasing the battery life of mobile phones with something you studied about chemical reactions in A level Chemistry
- the design of a new building with your study of forces in physics
- the need for engineers to be good communicators with your role as your class representative at school
- the follow-up research you did on wind turbines with an on-site visit on a school trip
- a news story on a new aircraft design with an article in the *New Scientist* about composite materials.

Work experience

Work experience is important as it demonstrates a commitment to the subject outside the classroom. Remember to include any experience, paid or voluntary. If you have had relevant work experience, mention it on your form. Explain concisely what your job entailed and what you got out of the whole experience. Even if you have not been able to get work experience, if you have spoken to anyone in engineering about their job it is worth mentioning as all this information builds up a picture of someone who is keen and has done some research. See Chapter 2 for further information.

How to get started on the personal statement

A good strategy is to start by making lists of anything that you think is relevant to your application. Then begin to organise them into sections. Your personal statement could include some of the following points.

I first became interested in engineering because …

- I read an article in a newspaper about …
- I read the book '…'
- I saw a piece on the news about …
- of my work experience
- of my father's job
- of something I have enjoyed studying at school.

I have investigated engineering by ...

- reading books
- reading the *New Scientist*
- reading the Royal Academy of Engineering website
- work experience
- going to a public lecture at a university
- discussing engineering with an engineer
- downloading a podcast of a university lecture on iTunes U.

From my work experience I learned ...

- that the qualities necessary to become a successful engineer are ...
- how the theory we study at A level is applied practically
- the importance of communication skills/problem-solving/leadership ...

Other points to include:

- a particular A level topic is useful because ...
- my part-time job is useful because ...
- my role as rugby captain has taught me ...
- being leader of the school orchestra has taught me ... (or lead in the school play, or ...)
- during my gap year I will be ...
- I was awarded first prize for ...

Only when you have the ideas listed should you start to write full sentences and to link the points.

Many people find that the best way to make it fit the available space is to start with a statement that is perhaps double the allowed word count but that incorporates everything you think is relevant, and then start to edit it down by:

1. removing any superfluous adjectives or complicated sentence structures, e.g. 'I was lucky enough to attend a stimulating talk on the exciting topic of wind turbines delivered by the renowned engineer Richard Martin, which inspired me to read xxx', could more succinctly be written 'I read xxx after hearing a talk on wind turbines by Richard Martin'
2. removing any passages that are not about you. Often, I read personal statements that try to impress the admissions tutors by explaining in detail engineering concepts that they have read about. The chances are that the person reading the statement is far more familiar with the topic than you are so will skip this section completely and so you have wasted valuable space which could have been used to tell him/her more about you
3. removing sentences that state the obvious: 'I chose physics and mathematics at A level because these will be useful when I study engineering'

4. removing passages that contain nothing factual, particularly closing statements such as 'I believe that my commitment to engineering coupled with my academic achievements and ability to work hard will make me an asset to your esteemed institution.'

Once you have gone through this process you should be getting near a personal statement that is around the right length. Than you can fine-tune it so that it fits exactly. A useful tip is to ask your referee to include anything that you have had to discard in his/her reference. So, if you have run out of space and have to omit information about, say, your role in organising a school event, then the referee can incorporate this into the Reference section of the form.

Things to avoid at all costs

- Boasting about your achievements.
- Telling the admissions tutor why he/she should offer you a place.
- Quotes from famous engineers, scientists or philosophers.
- Listing your examination results or predicted grades (these appear elsewhere on the UCAS form).
- Adapting or copying personal statements that you have found on the internet or in books (including this one!).
- Spelling mistakes or grammatical errors.

First impressions

Engineers tend to be practical people who can convey complicated information in ways that are easy to understand. They are problem-solvers who can get to the heart of an issue, discarding information that is not relevant to the issue they are dealing with. They understand the importance of turning ideas into solutions that have a benefit to others. Bear this in mind when writing the personal statement. Make the statement easy to read by using simple sentences, adding spaces between paragraphs, and sticking to the point. Give it a coherent structure and a logical flow. Remove anything that does not have a practical use. Treat the statement as you would an engineering project. Its purpose is to get you the offer at your chosen university.

Language

It is important that you use succinct language in your personal statement and make every word count. Remember that you are limited on the number of characters you may use so it is important not to use up this vital space with superfluous language. Keep it as simple and clear as you can, rather than using overcomplicated language in an effort to impress.

For example:

> *'I was privileged to be able to undertake an internship with a well-known engineering company where I was able to see the benefit of having the ability to be confident with information technology'* – approximately 200 characters.

Could be rewritten as:

> *'My three weeks' work placement at Rolls-Royce showed me the importance of being proficient in using spreadsheets'* – approximately 110 characters.

Similarly:

> *'I was honoured to be chosen to play the lead role in my most recent school drama production, and interacting with the producer and the rest of the cast involved a significant degree of communication and teamwork'* – approximately 210 characters.

Could be rewritten as:

> *'Performing as the lead in my school play taught me to work and communicate effectively with others'* – approximately 100 characters.

Phrases to avoid

- 'It was an honour to …'
- 'I was privileged to …'
- 'From an early age …'
- 'For as long as I can remember, I have dreamt of …'

'Writing the personal statement

- **Accuracy of spelling and grammar.** The current jobs market is such that companies are using software that automatically rejects applicants after three spelling mistakes. So it is important to demonstrate this attention to detail in a personal statement. The statement should not have any mistakes at all when it is finally polished and submitted. This probably means it will have been through 20 draft versions. It conveys a lot to an admissions tutor when an applicant has made a spelling or grammatical error, or has doubled or missed out words. If engineers don't pay attention to detail, there could be serious repercussions in most engineering disciplines, including safety issues.

- **Written communication.** Most practising engineers spend a lot of their daily working life writing good English in technical reports. Most applicants study mathematics, physics and chemistry and hence do not have formal opportunities to improve their written English. So, any evidence that an applicant has attempted to broaden their reading beyond textbooks or involved themselves in extended writing will be useful.
- **Structure.** Some personal statements are chaotic and rambling. Good ones have obviously been planned and tell a story about the applicant in a logical order. It is important for an applicant to realise how much time it takes to prepare a statement.
- **Peer review.** An essential step in constructing a personal statement is to let a close friend look at it. Frequently they will notice some important aspect of your life that you have forgotten to mention because it is so obvious.
- **The personal statement can steer the interview.** Some universities use the personal statement as a basis for allocating applicants to staff for interviews. So it is important to only include things that you want to discuss in these circumstances. So, including things because you think it will look good, but not knowing much about them, is a bad idea.
- **The personal statement must be personal.** It is usually quite obvious if an applicant is trying to guess what the reader wants. It is far better to write from the heart, demonstrating passion for the subject. So many personal statements look the same. So, when an admissions tutor reads one with some interesting element that nobody else has mentioned it livens it up. Start by thinking of five things that could only be included in *your* personal statement; in other words, make sure it is personal not formulaic.'

Lawrence Coates, Professor of Engineering,
University of East Anglia

Sample personal statements

The following examples of personal statements have been contributed by students who were successful in gaining offers from their chosen universities.

Please remember that these are personal statements; that is, they reflect the experiences and ambitions of the students who wrote them – so **do not attempt to copy them or adapt them for your own use**.

Personal statement 1

I became interested in engineering indirectly through my passion for cycling. I have competed in many amateur cycling events and competitions and I noticed that the most successful competitors were adept at doing their own modifications and repairs to their bikes. This became a serious interest after the Beijing Olympics following the success of the British cycling team in the track events and their 'marginal gains' philosophy – that every minute advantage that could be gained from the design of the bike or the helmet could add up to fractions of a second advantage over the other competitors, often the difference between a gold medal and no medal. This in turn started my interest in the use of materials and in materials engineering.

I work at weekends in my local bike shop (who also sponsor me in events) and build bikes from scratch to the requirements of serious cyclists. I also get to repair a range of machines that utilise interesting materials (including a bamboo road bike), and can appreciate the advantages and disadvantages of, for example, carbon frames compared to aluminium frames. My A level Physics and Mathematics are also useful in assessing and understanding things like the best cycling position for the rider (centre of mass) or disc brakes (energy conservation). I chose French as my third subject because learning a language will help me to work overseas at some point in the future, and it also stretches my academic abilities as I find it much harder than the sciences.

I attended a Headstart course this summer and found it helpful in understanding what studying engineering would be like, and also in meeting like-minded students who have interests in different areas of engineering. It also showed me that I was not the only female who wanted to be an engineer!

My EPQ is about the use of materials and, in particular, the efficiency of machines. I have built a simple hydraulic lifting system and have experimented with using different liquids and cords. Doing an EPQ has been, I believe, very beneficial in terms of making me think analytically and in being able to solve problems.

To keep up to date with engineering developments I read *New Scientist* and the science and technology section of the BBC website, and also many cycling magazines.

Other than cycling, I enjoy music and am the percussionist in the school orchestra and jazz band. I started off playing the drums but have expanded into using a whole range of percussion instruments, some of which I have invented and built myself. This has taught me how to work with others and also how to organise my time so that I can study effectively as well. This will be vital at university. I am on the school council and mentor younger students who have difficulty with mathematics. At weekends, when I am not competing, I work in my local Age Concern charity shop, sorting out clothes and working on the till. This has helped me to be able to communicate with a variety of people, old and young, which will help me when I work in the professional environment of engineering.

For my gap year I am going to undertake a two-month cycling tour of Europe with two of my friends. To fund this, I will continue to work in the bike shop and will also get a job in a local supermarket. My aim is also to build up my own part-time business building bespoke bikes from old bike parts and selling them. I believe that the gap year will make me a more focused and reflective student once I start my university course.

Personal statement 2

Since my early childhood I have always been fascinated with how things work. I remember being so curious about how my battery-powered toy car could move on its own that I had to open it up and try to re-assemble it after understanding how it works. To me, 'engineering' was the science that could create all the things around me. I thought pretty much all complex devices had two main aspects, an electric and a moving aspect, designed by electrical and mechanical engineers respectively. Mechanical engineering is, in fact, at the core of most of the devices that we use, ranging from children's toys, to planes and robots. My interest in mechanical engineering therefore stems from my childhood.

In June 2015, I underwent complex knee surgery to repair my torn meniscus by sewing the tear together. The best treatment available at the moment is a human tissue implant, which is very rare because it needs to be taken from a corpse yet is not very effective. I have been wondering why there is no artificial meniscus. What is it about an artificial meniscus that scientists cannot get right? Recently, I read an article that scientists from Imperial College London have managed to create an artificial meniscus, along with the necessary instruments for inserting it into the knee. This was a collaboration between the scientists from the mechanical and biomedical departments of the university. The hard thing was actually not to find a material similar to the meniscus tissue, but to get the mechanics of it, such as impact distribution right. It is fascinating to see how mechanical engineering has managed to influence and improve our everyday life, from our cars to the small and crucial components within our body. The beauty is in applying fundamental maths and physics laws to create such complex devices.

Mechanical engineering is the applied science of forces and movement, with engineers applying fundamental maths and physics laws to create and build mechanical devices that we use every day, fundamentals I have been learning during my A level studies and by reading around the course. I am interested in how maths reveals parallels between separate branches of physics. Kinematics, for example, is the branch of classical mechanics which describes the motion of points, bodies and systems of bodies without consideration of the causes of motion. The study of kinematics can be abstracted into purely mathematical functions. By using time as a parameter in geometry, mathematicians have developed a science of kinematic geometry. I am interested to continue my study in this at university.

I really enjoyed learning thermodynamics in A level Physics, concerned with heat and temperature and their relation to energy and work. Thermodynamic principles are used by mechanical engineers in the fields of heat transfer, thermofluids, and energy conversion. In order to ensure my degree path was correct, I undertook work experience at Iranian Compressed Air industrial Co. During my work experience and dealing with gas compressors of different types, I was fascinated by how very simple thermodynamic laws could have many different real life applications, ranging from household products, to large industrial ones. I was also drawn towards the business elements of the profession and am keen to extend my knowledge of business and management in order to widen my career choices in the future and gain useful skills for my entrepreneurial mindset.

Personal statement 3

I believe the modern world would be nothing without the breakthroughs engineers and scientists have made. I want to be a part of this process. I always pulled radios and watches apart, but it was aeroplanes that first drew me to science. I remember sitting by the window, grinning as the wing flaps flexed, while other passengers gripped the armrests and whispered prayers. I couldn't understand how anyone was afraid of flight. I am intrigued by how devices work, from simple electric motors to the principles of winglets on aeroplane wings, and by machines and systems, especially those that are theoretical or even fantastical.

I became interested in physics outside the classroom a couple of years ago. I've spent a lot of time reading ('New Scientist' and various websites) about mechanical processes, such as internal combustion, motorbikes' counter-steering, water saws, artificial gravity in space, and astro-engineering, especially Lagrange points and space elevators. I am following the efforts of the Japan Space Elevator Association to begin design and production. I find the progress being made in the field of nuclear fusion inspiring, especially after seeing a documentary on the development of HiPER. The use of lasers and mirrors was revelatory: the idea that a power station could create MeV of energy from isotopes in sea water seemed like science fiction but wasn't. I plan to visit the construction site of ITER next year to explore the practical side of nuclear physics and see how the theory I've learnt is applied in the real world. I heard a great Design London talk at Imperial College by Dick Powell about the connections between creative thinking, design and real world applications. Part of my desire to be an engineer comes from the solutions that science can provide – clean and sustainable energy, for example, especially in light of the BP crisis in the Gulf of Mexico; or safety and recovery issues highlighted by the stunning rescue of the Chilean miners. There is such a range of exciting projects; from the LHC at CERN, which I visited on a school trip, where I first began to understand that, while in theory nuclear physics happened on a small scale, the real world applications were massive; to the inventiveness of the 'barefoot engineers', like William Kamkwamba, who built a wind-powered generator, using bicycle parts, plastic pipes and simple motors.

My interest in RE and ancient cultures (the ethics and philosophy behind science) has broadened my interest in engineering by adding a sociological element. I have been lucky enough to travel widely, and seen several impressive engineering achievements; the Sydney Opera House and the Harbour Bridge are two I will never forget. Jorn Utzon's ingenious interlocking shells showed me how design and practical engineering are so interdependent – a life-altering moment. Recently travelling in Israel and witnessing the huge Jewish settlements erected in months rather than years was amazing. Trekking to the ancient city of Petra in

Jordan, carved out of mountains, made me understand that even 2,000 years ago the principles of engineering and design were just as vital and universal as they are now.

I was an intern at the design company Public Creative, indexing, using Photoshop and developing my people skills. In this gap year I have planned work experience for WSP Group, as well as at Meggitt and Classic Aero Engineering, restoring a Spitfire and a Hurricane.

Music is another passion: playing saxophone in the school jazz band for 10 years meant lots of concerts at school, the Eisteddfod, in London and Paris. I am a good listener and team player. I enjoyed volunteering at a local primary school and helping children to read.

I am fascinated by the principles and practice of engineering and look forward to the challenge and excitement that university life will bring.

Personal statement 4

A Japanese female racing engineer – that is something you never really hear of even in the motorsport industry. I always get asked why a girl like me is interested in motorsport, and one of the main reasons is because my family raised me without fixed ideas, such as that girls should do 'girly' things. Since my grandfather loves cars, I would follow him to the dealership just to see new models and go for a test drive. I've always liked cars, so it was clear to me that I wanted to study engineering at university and eventually get a job in that field. This led me to take physics and mathematics at higher level for the IBDP.

In Year 10, I got to listen to the internship experiences of the older students. I was fascinated by the idea of doing an internship both for my UCAS and my future. By the end of year 10, I was able to find an internship that changed my perspective entirely. During the summer of 2015, I had a chance to spend time at a Japanese Toyota/Lexus based racing team called TOM'S. I helped the racing engineers with the data processing from previous races and the preparation for the next, by preparing brake pads, changing the air pressure in the tyres, and timing practice pit stops. I also learned some CAD skills when I visited one of the factories where they made parts for the race cars. I was subsequently invited to watch the Suzuka 1000km race from the pit – a race TOM'S won. It was significantly different from watching the race on a screen. I was able to really understand what it is like to be a racing engineer and realised that they need to be ready for anything; to be precise, calm, and collected. Being able to interact with the team directly and watch them do their work helped me grasp the reality of life as a racing engineer, the great things they do and how tough it can be. It was the first time I felt a strong sense of belonging and passion, and I will never forget every second I spent there.

Since the internship, I've been fascinated with motorsport, and physics has become my favourite class in school. My internal assessments for physics, chemistry, and business were all related to automotive engineering, such as investigations into aluminium alloys, the combustion of biodiesels based on different vegetable oils, and a study of Toyota's autonomous car technology. My extended essay (EE) was heavily inspired by my internship, as I conducted an experiment on the NACA duct, with a handmade wind tunnel. My EE required commitment, patience, and the ability to work around problems. The process of completing the EE also helped me imagine what it would be like to study in university and to be a racing engineer, leaving me even more excited about the future.

Not only have I explored the technical skills needed to be a racing engineer; I have also been able to develop the necessary interpersonal skills. In February 2016, I was selected as the student leader of the International Student Forum, a major conference held by our school. I was able to take the role of leading in the management team for the period before and during the conference. This job was a heavy responsibility, but through this experience I cultivated the communicative, decision-making, and cooperative skills that are essential in modern motorsport engineering teams. Even though I've only lived in Kyoto, I have studied at an international school and gone through the IB programme in high school, which has made me a very open-minded person.

So far my family, friends, and most of the people I meet are supportive of my dream. I'm determined to study in the UK, the home of motorsport, and become a racing engineer. I hope in the future I will be able to show what a Japanese female engineer is capable of.

Personal statement 5

I decided on becoming an engineer at quite a late stage in my education. My favourite subject at school was physics, and in particular theoretical physics, and this is what I intended to study at university. As part of my IB programme, we are required to work on an extended essay and mine was based around semiconductors. It was through this that I became more interested in the role of semiconductors in practical situations rather than how they work or how different types can be combined to create logic gates.

To investigate this further, I contacted the Institute of Electrical and Electronics Engineers and they introduced me to the Try Engineering project. Through this I was able to make contact with engineers and engineering students to find out more about what studying engineering and becoming an engineer would entail. And then I was hooked! Through one of these contacts I was able to do some work experience last summer, in a small start-up company specialising in GPS products, and I found this very exciting, even though most of the technical aspects were way beyond my understanding. But I did learn that engineers have to be able to be practical people because they are producing things that other people will hopefully want to buy. Following this experience, I read 'Electronics: The Life Story of a Technology' by Morton and Gabriel, which explained the history of the development of the semiconductor and how developments were always accelerated by practical requirements.

My favourite topics in physics are those related to electricity and electronics, and I enjoy the practical side of the subject very much. As part of a project organised by the school, I created a light-meter using an LDR and different colour LEDs to indicate whether the light level was suitable for taking photographs without a tripod. My strength in mathematics is also going to be very important as my research has shown me that mathematics courses are an integral part of an engineering degree. Other than mathematics and physics, I am also studying economics and art. Economics is very helpful because as well as being a subject where theory has to be applied to, and developed from, real-life events, it also shows me that to be successful, a product has to be priced correctly and marketed effectively. Art is helpful in helping me to think 'outside the box' and in a creative way, as well as a chance to use my hands to create things.

When I am not studying, I enjoy sport. I am captain of my school's soccer team and I also enjoy swimming, badminton and martial arts. I play the saxophone in the school band which helps me to understand the importance of practice and commitment, and how to work effectively with others. I am on the school's charity committee and we recently held a 'Dragon's Den' event where we offered a small sum of money to the person in the school who had the best idea of using it to make something to sell to raise money for our chosen charity (a school project in a poor region of Thailand). It was interesting to do this because we had to decide not which was the best idea, but which was the most practical and most likely to raise money.

I am confident that engineering is the right career for me, and that I have a lot to offer my chosen university both academically and socially.

General tips for completing your UCAS application

- Before submitting it, also ensure you check your application through very carefully for careless errors that are harder to see on screen.
- Keep a copy of your UCAS application so you can remind yourself what you wrote, prior to an interview.
- Ensure that you have actually done all the things you mentioned in the statement by the time you are interviewed.
- Research the full course content, not just the first year.
- Research the entry requirements.
- Ask your teachers for your grade predictions.
- Ensure your personal statement is directed at the courses you are applying for.
- Include lots of detail in the personal statement.
- Get someone else to proofread your personal statement.
- Illustrate your points with examples and evidence.
- Do not waste valuable space in the personal statement – make every word count.

6 | Succeeding at interview

Not all universities interview candidates, but it is likely that at least one of your five choices will do so, so you need to be prepared. Your interview will decide whether you will be offered a place or not. The information on the UCAS application will have been the basis on which the decision to interview you was made, but a good UCAS form cannot help after a poor interview. So prepare thoroughly. Here are some important points to consider.

- If you interview well, and you subsequently narrowly miss the grades that you need to take the place, you may still be offered the place.
- Interviews are normally conducted in an informal and relaxed manner, the purpose of which is to allow you to talk about your interests and suitability for the course.
- Think about the impression you will make – think about your body language, eye contact and communication skills.
- Go into the interview with a mental checklist of what points you wish to mention and try to steer the interview to address these (see below).
- Interviewers are less interested in investigating your subject knowledge than in looking at how suitable and committed you are for their course. So, evidence of research and appropriate qualities such as analytical or problem-solving skills are important elements of a successful interview.
- Remember, your future teacher or lecturer might be among the people interviewing you. Enthusiasm and a genuine commitment to your subject are extremely important attitudes to convey.
- An ability to think on your feet is vital – engineering is about problem solving. Don't try to memorise potential answers or responses, but be prepared to expand on things you have mentioned in the personal statement.
- Important preparation includes re-reading your UCAS personal statement. Never include anything in your UCAS application that you are not prepared to speak about at greater length or in more detail at the interview.
- Questions may well be asked on your extracurricular activities. The interviewer may do so either to put you at your ease or to find out about the sort of personal qualities you possess; therefore your answers should be thorough and enthusiastic.

- At the end of the interview, you may be asked if you have any questions. Often, this is simply a polite way of ending the interview, so do not feel that you need to ask anything. Just say 'Thank you, but all my questions were answered during the introductory lecture today and by the students who showed us around. If I think of anything I will contact the admissions department.'
- Finally, smile, thank them and shake hands. Above all, convey your enthusiasm so that they will remember you at the end of a long day of interviews.

'Within the Department of EEE, we actively encourage all applicants that are expected to meet our very high entry standards to attend their interview afternoon. The reason is not to test their academic ability, but to ensure that they have what it takes to succeed at Imperial in their studies. Perhaps more importantly, we want our high-calibre students to be happy during their stay at Imperial and so it is important to know that they want to study EEE for the right reasons. To this end, during their interview, we are looking for a student who is 'switched on' (as opposed to an applicant having been primed for the occasion) and genuinely keen to pursue a career in EEE (it is not uncommon for parents to push their children into this subject). We ask each applicant a few technical questions, of increasing difficulty, to gauge their limits of mathematics and physics; not necessarily related to EEE. In addition, we are looking for a confident, inquisitive mind, as well as signs of weakness areas. Since places are limited, we try to find signs that the applicant will have a long-term commitment to academic life, as well as indicators that suggest that they will help to enrich the lives of those around them.'

Dr Stepan Lucyszyn, Undergraduate Admissions Tutor,
Department of Electrical and Electronic Engineering (EEE),
Imperial College London

Preparing for an interview

If you are called to interview, it is advisable that you reread your personal statement to remind yourself of what you wrote, so that you can be ready to answer questions on it at interview. Preparation for an interview should be also an intensification of the work you are already doing outside class for your A level courses. Interviewers will be looking for evidence of an academic interest and commitment that extends beyond the classroom. They will also be looking for an ability to apply the theories and methods that you have been learning in your A level courses to the real world.

Essentially, the interview is a chance for you to demonstrate knowledge of, commitment to and enthusiasm for engineering. The only way to do this is by trying to be as well informed as you can be. Interviewers will want to know your reasons for wishing to study engineering and the best way to demonstrate this is with examples of things you have seen, read about or researched. Later in the chapter there is a section on current issues that you can use to kick-start your reading.

Newspapers and magazines

Before your interview it is vital that you are aware of current affairs that relate to the course for which you are being interviewed. *New Scientist* will give you a good grasp of scientific and engineering developments, as will reading the science sections of the broadsheet newspapers. You should also keep up to date with current affairs in general.

Magazines can be an important source of comment on current issues and deeper analysis. There are many specialist engineering publications, such as *The Engineer*, *Structure Magazine* and *Aviation Week*. Further details can be found in Chapter 11.

Television and radio

It is also important to watch or listen to the news every day, again paying particular attention to news about scientific and engineering issues. Documentaries and programmes about engineering projects can be enormously helpful in showing how what you are studying is applied to actual situations and events. Keep an eye on the television schedules for programmes or series on anything related to your field of interest, which could range from those aimed at a wide audience (*Grand Designs*, *James May: The Reassembler*) to more factual programmes such as BBC's *Horizon* and programmes on engineering projects on the National Geographic Channel and Discovery Channel. BBC Radio 4's series *Inside Science*, *Frontiers* and *Material World* are also very useful.

The internet

A wealth of easily accessible, continually updated and useful information is, it goes without saying, available on the internet. Given the ease with which information can be accessed, there is really no excuse for not being able to keep up to date with relevant current issues. Radio programmes can be downloaded as podcasts and listened to at times convenient to you; the BBC's iPlayer gives access to current affairs and documentary programmes; iTunes U gives free access to thousands of lectures and presentations from universities around the world;

newspapers can be read online ... the list is endless. In this age of information overload, anyone who is serious about keeping abreast of current issues (or wants to be seen as being serious) has unlimited opportunities to do so. Thus, an interviewer is not going to be impressed with a student who claims that he or she has been too busy to know what is happening in his or her chosen areas of interest.

- Subscribe to podcasts and download them regularly. BBC podcasts, which are free, include *Inside Science* and *Material World*.
- Check online news websites every day to read the latest news stories.
- If you cannot buy a newspaper every day, look at an online version, for example www.theguardian.com.

Examples of your areas of interest

One way to make an interview a success is to illustrate the points you are making with examples. It is also easier to talk about something you know about rather than trying to talk in general terms. And if the examples you use are interesting, the interviewer may well want to talk about them rather than ask you the next question on his or her list. But remember, this will only work if you have done your research beforehand. There is nothing worse in an interview than a conversation along the lines of:

You: One of the things that inspired me to study civil engineering was a journey with my parents up the east coast of England, when we crossed the Humber Bridge, the first suspension bridge I had ever seen at first hand.
Interviewer: I see. Can you tell me something about the reasons for building a suspension bridge there rather than a beam bridge?
You: Sorry, I don't know.

A better answer would have been:

You: The Humber estuary is used for shipping, and because of the width of the estuary, a traditional beam or cantilever structure would not have been able to span the space between the banks. Also, suspension bridges have some flexibility and that area is prone to high winds.

The interviewer might then have gone on to discuss the types of forces that are present in a suspension bridge, and about suitable materials, all of which you would be familiar with because you had anticipated this response and had prepared for it.

Here are some ideas to use as examples to illustrate points you want to make:

- civil engineering: names and construction details of bridges or a new airport

- structural engineering: how the use of new materials helped in the building of a new skyscraper or office block
- mechanical engineering: examples of machines that you can discuss (cars, aircraft, wind turbines, etc.)
- chemical engineering: an industrial chemical process
- biomedical engineering: a medical breakthrough that was developed by engineers (such as computerised tomography (CT) scanners)
- electronic engineering: the background on the development of, for example, solid state memory for computers.

The interview

Interview questions are likely to test your knowledge of engineering projects and developments in the real world, since, unlike some theoretical science subjects, engineering is a practical subject aimed at making the world a better place. It is important that your answers are delivered in appropriate language. You will impress interviewers with fluent use of precise technical terms, and thus detailed knowledge of the definitions of words and phrases used in engineering is essential. Potential electrical engineers need to know the technical and microscopic differences between semiconductors and insulators, and to be able to differentiate clearly between electric charge and current; and if you are interested in materials or civil engineering, you need to use words such as stress, strain, elasticity, strength, toughness and stiffness with their scientific, rather than their everyday, meanings.

One popular question is to ask which topics you have enjoyed studying at school. Be prepared for this by doing some revision so that you are not desperately trying to remember details from things you studied a year ago. Try to talk about something that is closely linked to engineering.

You may be asked about your future career plans. If you are applying for a particular field of engineering then your future area of specialty will be apparent, but you may be applying for a general engineering course, or one where you specialise only in the second year, so be prepared to talk about your plans. You may have ideas about where you want to work, in a big company or possibly overseas. It is a good idea to relate your possible plans to research you have done, or to your work experience. It is also a good idea to demonstrate a knowledge of how you gain Chartered Engineer status (see page 111).

You may be asked questions that appear to want your opinion on a recent development or issue. This type of question is asked to see whether you have been thinking about engineering issues, or whether you have been keeping up to date with current issues. Ultimately, the interviewer is not really interested in your opinion but in your ability to formulate arguments and your interest in the field.

Practice of interview situations, like most other things in life, will make you better prepared, less nervous and more confident. Arrange mock interviews with teachers, friends of your family, or with careers advisers.

Some things that can adversely affect the interview include:

- arriving late and flustered
- being unable to talk about things mentioned in the personal statement
- not listening to the question carefully before answering
- interrupting the interviewer.

Student experience

'My interview at Cambridge was quite different to the other university interviews I faced. Firstly, there were two interviews, both conducted by a panel of two members of the Engineering Faculty. The questions asked were nearly all technical in nature, with only one or two being personal questions or about my extra-curricular activities.

'It is crucial that you have revised your A level subjects before the interview because you will be expected to apply techniques and principles learnt during your A levels in your interview. My interview included questions that I did not initially understand but thankfully the interviewers do not expect you to know everything, rather they are looking for people who are teachable and learn quickly. During the interview, I was guided through questions to help me reach the answer and the interviewers were looking to see how quickly I picked things up and to see how I thought things through – this is why it is better to voice your thoughts than to think in silence during the interview. I was asked to derive equations about the lift created by aircraft wings. I left my final interview thinking that I hadn't done well but it's often the case that you can feel like you've got a lot of questions wrong because the interviewer's task is to push you to the limit of your understanding and see how you handle it. It's almost impossible to predict how well a Cambridge interview has gone, so don't worry if you don't feel you did well.'

Jordan Massiah, Magdalene College, University of Cambridge, MEng Engineering

Student experience

'Regarding interviews for mechanical engineering, I would recommend revising calculus, graph sketching and physics, such as mechanics and electricity. I used a site called www.i-want-to-study-engineering.org which has many questions and solutions for many engineering type interview questions. Also, I found that revising from my school textbook was useful for brushing up on old techniques. My interview experiences were as follows:

'First interview:

'I was given a situation where a hopper was filled with sand and placed on some scales. The task was to draw the graph of the reading the scales would show if you let the sand fall out of the hopper and onto the scales, from the time the first grain of sand started to fall until the last one fell. The next question was about a cube hanging from one of its vertices and the task was to find the shape you would get if you cut horizontally across it from different heights. (No drawing was allowed.)

'Second interview:

'I sat a one-hour paper before the interview and we went over it with the interviewer.

'To start, there were some graphs to draw:

$y=(x+1)/(x-1)$

$y=(x+1)^2/(x-1)$

Then there were questions about circuits with resistors in parallel and in series. In a circuit with a cell of voltage V, there are two resistors in parallel to each other, which are then connected in series to another pair of resistors which are also in parallel to each other. The task was to find the power dissipated by one of the resistors.'

D.S. Barreto Romero, University College London,
MEng Mechanical Engineering with Business Finance

50 sample interview questions

A. About you

1. What first started your interest in engineering?
2. Why do you want to be an engineer?
3. What qualities does it take to be a successful engineer?
4. What have you done to investigate engineering as a course?
5. What have you done to investigate engineering as a career?
6. What field of engineering particularly interests you?
7. What did you learn from your work experience at 'X Enterprise'?

B. Your research into engineering

8. As an engineer you will be involved in creating commercially viable products. Can you discuss a situation where the commercial pressures are in conflict with environmental issues?
9. Give me a very brief outline of the key engineering developments of the twentieth/twenty-first century.
10. What do you consider to be the most significant engineering project in history?
11. Do you have an engineering hero/heroine?
12. What do you think is the difference between science and engineering?

C. Academic questions

13. How does the structure of a metal determine its properties?
14. Why does the molecular structure of wood make it suitable for some building projects but not others?
15. What is the difference between a 'tough' material and a 'strong' material?
16. What do we mean by potential difference?
17. What is the difference between charge and current?
18. Why are the concrete girders used to construct buildings 'T' shaped in cross-section?
19. How does the internet work?
20. What is meant by conservation of energy?
21. What is meant by a 'cantilever'? How do the stresses on a bridge using cantilevers differ from a bridge using a beam to span the distance between two supports?
22. Which parts of the electromagnetic spectrum can humans detect?
23. Can you explain what is meant by 'proof by induction'?
24. Can a scientific theory ever be proved?
25. What is a semiconductor?

D. Unexpected questions (designed to test your ability to think creatively)

26. Murphy's law says that whatever can go wrong will go wrong; for example, if you drop a piece of bread that has jam on one side on the floor, it will always fall with the jam side down. How would you go about verifying Murphy's law?
27. Engineers cause more problems than they solve: do you agree?
28. What is a machine?
29. What is a computer?
30. What is a bit and what is a byte?
31. What is 27 in binary? What would it be in a system based on the number 4 rather than 2? Or in a number system based on the number 9?
32. Why is nanotechnology so called?
33. How does a car engine work?
34. Why is the distance that an electric vehicle can travel so small compared with a petrol-fuelled vehicle?
35. Is there any such thing as a 'new' invention?
36. How does an aircraft stay in the air when it is more dense than air?
37. What is meant by the words 'digital' and 'analogue' when describing communication systems such as TV signals?
38. People describe new inventions as being the most significant since 'the invention of the wheel'. How do you think the wheel was invented?
39. People describe a good idea as being 'the best thing since sliced bread' – what are the advantages and disadvantages of sliced bread?
40. Why does a bicycle have gears?
41. How does the car braking system work?
42. What limits the maximum height of a proposed new office development?
43. What is a robot?
44. An architect designs a 50-storey office building. How might an engineer test whether it is safe to build it?
45. An architect designs a steel and glass bridge. How might an engineer test whether it is safe to build it?
46. A designer creates a model for a new type of passenger aircraft. How might an engineer test whether it is safe to build it?
47. What is biotechnology?
48. What will replace fossil fuels when they run out?
49. What do we mean by 'alternative' energy sources?
50. Describe this (showing the student an everyday object – a chair, a frying pan, a light bulb, a shoe, a watch) from an engineering perspective.

How to answer interview questions

Introductory questions

Why have you chosen to apply here?

The interviewer will need to be reassured that you have done your research, and that you are applying to the university for the right reasons, rather than because your friend tells you that the social life at that particular university is excellent.

Your answer should, if possible, include the following points:

- first-hand knowledge of the university, for example you came to an open day or you have spoken to students who have studied there; if you cannot visit the university, then at least try to discuss the institution with current or ex-students (many university websites have links to current students who can answer your questions directly)
- detailed knowledge of the course and why it is attractive to you, or how it links to your future career plans. The course might, for example, offer the chance to learn a language as one of the options in year 2, and you could mention this as being something that will help you to work overseas. Or it might offer work placements or the chance to spend a period of time at an overseas university.

Why do you want to be an engineer?

(This question, or a variant on it – What have you done to investigate engineering? When did you decide that engineering was the right course for you? – will almost certainly be asked. It would be considered by the interviewers to be a gentle introduction to the interview because they will assume that you have thought about this, and anticipated it being asked.)

Your answer should include an indication of how your interest started (for example taking apart a radio, building a model out of Lego, something you were taught in a science lesson) and lead on to things you have done to investigate engineering. This would, ideally, involve work experience or an engineering lecture you went to. You could end up by talking about a particular area of interest (mechanical engineering, civil engineering) or a project that interests you (a building, a machine?) and possible plans for your future career.

General questions about engineering

What qualities should an engineer possess?

(Variants on this question might include 'From your work experience, what did you learn about what it takes to be a successful engineer?')

Points you might raise could include: mathematical ability, logic, analytical and problem-solving skills and curiosity. But it is important to expand on these rather than simply list them. Explain, ideally using an example to illustrate what you are saying, why you think that this quality is important. Examples can be drawn from your work experience, your wider reading, or a lecture.

Here is an example:

> *'The ability to solve problems is very important. I really became aware of this when I was doing my work experience at a local engineering company. They were making low-voltage lamps for use in recessed lighting fittings in houses and offices, but in one particular building, the lamps kept blowing. In the end, one of the engineers decided that the problem must have been in the transformer rather than the lamp itself, and so he looked at where the transformers for each lamp were situated. It turned out that they were short-circuiting because the cavity above the false ceiling was damp.'*

What does a mechanical engineer do?

(Variants on this could include asking for definitions of engineering, science or technology.)

Again, try to illustrate your answer with an example or the details of a conversation that you had with an engineer:

> *'Mechanical engineers work with machines. But I know from the work experience that I did at my local garage that understanding about the mechanical aspects of, for example, a car engine is only a small part of it. You need to have a good knowledge of electricity and confidence with computers, since a lot of the trouble-shooting was done using electronic equipment.'*

Questions designed to assess your clarity of thought

You may be given an open-ended question about something you have already studied. The point of this type of question is not so much to test your knowledge or academic level (because this will be clear in the grade predictions and exam results on the UCAS application) but to see if you can think logically and in a structured way. So, your answer should really be an exercise in 'thinking aloud'; that is, talking the interviewer through the steps to your final answer. An example is given below.

Why are metals so useful to engineers?

You could start from first principles by describing the microscopic structure of a metal. This shows that you can approach problems in a logical way while also giving you some time to think about where your answer is going: 'A metallic structure consists of a lattice of positive ions surrounded by a "sea" of delocalised electrons. It is this structure that gives metals their useful properties.'

You might then go on to look at a number of properties in detail: 'The most obvious properties this gives metals are good electrical and thermal conductivity. Electrical conduction is through the flow of electrons through the lattice, and since they are not attached to any particular atom, they can move freely. Metals are also good conductors of heat because the electrons are able to transfer energy as they move in addition to the vibrations of the lattice.'

You could then move on to other properties that make metals so useful, describing each one in turn. You would probably include malleability, and the use of physical processes to alter the strength, stiffness or toughness of a metal to suit its intended usage.

Questions that assess your ability to analyse or to solve problems

You may well be confronted by a question about a situation that you have not covered in your studies. Don't worry. The interviewer will know that this is a new area for you. What he or she is looking for is not for you to immediately give them the correct answer, but rather how you can take things that you know and apply them to new situations.

Applicants for one university were asked 'What percentage of the world's water is in one cow?' Of course, no one (including the person who asked the question) knows the answer to this. What they were interested in, as discussed in the previous example, was in the candidate's ability to approach a problem from first principles and to arrive at an answer in a logical and structured way. So, an answer of 'I don't know' would not be very useful to your chances of being offered a place.

A better answer might start with: 'Well, I suppose I might begin with trying to estimate how much water there is on Earth. I know from my physics what the radius of the Earth is, and so I could work out its surface area. I could then make an assumption about the average depth of the oceans and work out their volume …' Rather than listen in silence, it is likely that the interviewer will help you by giving you hints or guiding you. But they can only do this if you explain every step.

When I talk to students about what they worry about when they are preparing for their interviews, they always say, 'What if I cannot answer a question?' And here is what I say to them.

- Interviewers are aware of the level you have studied to, and so will have a good idea of what you should know and may not know.

- Therefore, it is likely that any 'new' topic that you are confronted with at the interview has been asked to see how you think rather than what you know.
- Approach all questions from first principles, for example GCSE knowledge, and then build up your answer.
- If your answer requires you to draw a sketch or do a calculation, ask if you can use a piece of paper or the whiteboard in the interview room.
- Don't be afraid to ask for help, but do this by asking for comments on what you think is the right approach: 'I think I would start by looking at the forces on the body – is this right?'

Questions to show your interest in engineering

Anyone can say that they are interested in engineering, but by applying to study engineering at university, you are embarking not just on a short period of study but on your future career as well. An interviewer (who is almost certainly an engineer) will want to be reassured that you are serious enough about the profession to keep up to date with developments and engineering issues. So, questions such as 'Tell me about an engineering issue that you have read about recently' are designed to see if you keep abreast of current events. How do you ensure that you are prepared for such questions?

- Watch the television or listen to the radio news on a daily basis. Read the quality newspapers as often as you can, and keep a scrapbook of engineering-related stories.
- Check websites for engineering stories. A good starting point is the BBC website (www.bbc.com), which has a section on science and technology.
- Visit the news sections of the websites of the engineering institutions and professional bodies (see Chapter 11 for website addresses).
- Talk to engineers.
- If you can, go to public lectures at universities. Details can be found on the university websites.
- Watch podcasts of university lectures on iTunes U.
- Download or listen to radio programmes such as BBC Radio 4's *Material World*.

Steering the interview

There will be issues that you want to raise in the interview, things that will demonstrate your research, commitment and personal qualities. Rather than walking out of the interview disappointed that you did not have the opportunity to discuss these things, try to bring them into the conversation. For example, you may have been to a lecture on developments within the electronics industry at a local university one evening and you want to talk about this. There are likely to be many ways that you can do this. You might be asked why you want to be an engineer, and during your answer you could say 'and the thing that really convinced me that electronic engineering was the right career for me was listening to Professor Smith talking about nanotechnology at a lecture that I attended at Surrey University last month'.

In all probability, the interviewer will then ask you more about this, and you can then talk about something that you know about, rather than having to face questions on a topic with which you are less familiar.

Before you go to the interview, write down a list of things you want to talk about, and think of ways that you may be able to do so.

Current issues

As a potential university engineering student you need to demonstrate your interest by keeping up to date with current issues and developments. Engineering is an ever-evolving subject, with new materials, processes and products being developed every day. Just think about the rapid changes in the field of communications over the past 15 years, for example. Once you have identified your particular areas of interest, you need to keep researching and reading about them, perhaps keeping a scrapbook (either a physical one or on your computer) of news articles that are relevant to your chosen area of study, and using this as part of your preparation for an interview.

In this section, I have covered some broad topic areas that should be familiar to potential engineers. The summaries in this chapter are included to give you an illustration of the kind of events and other news items you should be reading about. In other words, they are a starting point for your own research, rather than being an easily accessible source of information to memorise prior to your interviews.

Engineering and materials

Engineering is about making practical things and so engineering projects require materials – that is obvious! But what factors determine which materials are suitable for a particular project? And how do engineers go about choosing between materials that may appear to have similar properties? To do this, engineers need to be able to describe the properties of the materials they are looking for. Physicists and engineers have to be very precise in the way they describe the properties of materials, and words that we use in a very general sense in our daily lives have very specific meanings for engineers. Words such as 'tough', and 'strong'; or 'elastic' and 'plastic'; words that we use in normal conversation and might appear to mean similar things have precise definitions when used in an engineering context. Prospective engineers need to learn the vocabulary of materials.

When we talk about properties, we mean how a particular material behaves when subjected to external factors – light, forces, heat, electricity, time. And a property that might be useful in one context, for example, stretching elastically, might be problematic in a different situation.

Some of the properties of materials that engineers are interested are:

- electrical conductivity
- thermal conductivity
- how they behave when subjected to forces
- how their behaviour changes over time
- resistance to corrosion
- how they are affected by heat or temperature
- how different materials can be joined
- resistance to chemicals or changing weather conditions
- toxicity
- aesthetic properties – will people want to buy or use the product?

Consider your smartphone. What materials are used and why? Electronic components rely on their ability to conduct (or not) electricity. The electronic circuits and microprocessors in your phone are built out of semiconductors. Semiconductors are somewhere in between electrical conductors and electrical insulators and the way that they conduct electricity depends on the materials used. And so electronic engineers are interested in the properties of semiconductors. The flow of electricity through your phone through the conducting parts of the circuits generates heat, and so the engineer is also concerned with minimising this and thus needs to understand the different metals that are available. Some of these are cheap, some expensive, some last a long time, some corrode, some are safe to be in contact with, some are toxic.

Your phone relies on a battery to supply the electricity that it needs to function. This battery contains chemicals. It needs to be able to be recharged regularly without deterioration of the chemicals, and to have a long lifetime. And what happens when the phone is defunct and has to be thrown away? What effect will these chemicals have on the environment? How safe are the batteries? If you were the engineer that developed the batteries in a recently available smartphone that appeared to have a tendency to catch fire and the company that developed it had to recall millions that had been sold, your future job prospects might not be that promising.

What does the designer want the phone to look like? Should it be available in different colours, or look metallic? And how resistant will it be to wear and tear or accidents? Some materials are shiny, some are transparent. Some are strong, some will crack when the phone is dropped.

Some get hot very quickly. Materials engineering is a field that answers questions like these.

Let's look at another example: building a new iconic skyscraper. An architect will create the initial design of the building, which will be based on aesthetic ideas, as well as thinking about what the building will be used for – 'form and function'. This would probably start as a rough sketch and then the architectural practice will work on more detailed designs. This is where the structural engineers will be involved. First and foremost, the building has to be structurally viable; that is, it shouldn't fall down. So the structural engineer will need to look at all aspects of the structure: will it be built around a central concrete core with the exterior fixed to this, or are the external walls going to be load-bearing? Concrete is very strong when subjected to compressive forces, but cracks easily when tensile (stretching) forces are applied. So if tensile forces are involved, then it has to be strengthened or toughened (see below) by adding other materials to reinforce it or pre-stress it. If the building is going to be tall, then how flexible does it need to be in strong winds? How will people get to the upper floors? What materials should be used for the lift cables? What about ventilation? What type of glass? A building with a glass exterior will get very hot because of the greenhouse effect.

A brief explanation of some of the commonly used terms that describe materials or their properties is given below. But you should read more about the properties of the materials that are used in the particular field of engineering that interests you.

Types of materials

Metals: the atomic structure of a metal is what gives metals their most useful properties. Because the electrons are relatively free to move through the fixed crystal lattice structure, metals are good conductors of electricity and heat. These 'delocalised' electrons also make metals malleable and shiny.

Polymers: made up from long chains of repeating units. These molecular chains can be arranged in more or less orderly patterns, or tangled up randomly. This means that polymers can usually be stretched and can be moulded into shapes. Polymers are chosen for particular uses by looking at their elastic or plastic behaviour, and by looking at how they react to heat, light or chemical exposure.

Amorphous: as the name suggests, the atoms and molecules are randomly arranged with no long range discernible order. Glass and ceramic materials are amorphous in structure.

Composites: composites are composed of different materials (or the same material in different orientations) and exhibit the best properties of each. Reinforced concrete incorporates steel rods within the structure so is strong in compression and also in tension.

Properties of materials

Strength: strength is a measure of how much stress a material can withstand (a similar measure to pressure) before breaking.

Toughness: not the same as strength. A tough material might or might not be strong, but it will not be brittle, that is, it won't break through cracks appearing and spreading through the material. So glass can be strong but is not tough (unless it is laminated or treated to improve its toughness), whilst a digestive biscuit is neither strong nor tough. A polythene bag might not be very strong but it is tough.

Stiffness: a measure of how easily the material deforms – bends or stretches or compresses – when forces are applied.

Brittleness: brittle materials tend not to stretch very much when a force is applied, but then suddenly break. Glass is a brittle material.

Elasticity: elastic materials deform when forces are applied, but then return to their original shape and size when the force is removed. Why is an elastic band so named? Because it exhibits elastic behaviour.

Plastic behaviour: the opposite of elastic. If the force is removed the material does not return to its original shape. Take a plastic supermarket bag and cut a thin strip from it and then stretch it. Initially it will not stretch much but when you apply a certain force it will suddenly be easy to stretch by a significant amount. But when you stop applying the force it will not go back to its original size. So a plastic bag is so called because it behaves plastically.

Malleability: metals are malleable. It means they can be hammered or pressed into new shapes.

Electrical conductivity: when a potential difference is applied to a conductor, how much current will flow? The higher the current, the higher the conductivity. Electrical conductivity may be a desirable or undesirable property. Electrical insulators are also required in any devices that use electricity. Semiconductors are materials that conduct electricity by an amount determined by how it is made or at what temperature it operates.

Thermal conductivity: a measure of how much heat flows in a material. As with electrical conductivity, both good and poor thermal conductors are useful to engineers. Cooling is important in many mechanical devices where heat is generated, and so good conductors are needed to remove the heat efficiently. But the handle of your kettle should be a good thermal insulator so you do not burn your hand.

Failure mechanisms: a material or component can fracture or stop being suitable for its chosen purpose in many different ways and from many different causes. They can crack, deform, melt, corrode, decay, suffer from metal fatigue, or stop being useful in many other ways. The causes of this could be defects in the molecular structure (metal fatigue is caused by dislocations within the structure), the application of too much force or heat or electric current, through chemical reactions, surface scratches or even by being joined to other materials using unsuitable adhesives or techniques.

Commercial issues

Someone has to pay for the things that engineers produce. Sometimes you pay through your taxes (roads, bridges, medical equipment), but in most cases the things that engineers make have to be useful or attractive enough for people or companies to pay for them. Two very similar, in terms of functionality, smartphones or cars may vary in price by a factor of four or five simply because of the materials used.

Engineering, ethics and the environment

Engineers are professionals and, as such, they have to behave in a professional manner, and ensure that they act ethically at all times. Engineers need to be trusted by the public. The work that engineers do affects millions of people – think of engineering projects that involve defence, power generation or the provision of clean water, for example.

The Royal Academy of Engineering and the Engineering Council sum up the obligations of engineers in their Statement of Ethical Principles:

1. Accuracy and rigour
2. Honesty and integrity
3. Respect for life, law and public good
4. Responsible leadership, listening and informing

Reproduced courtesy of the Royal Academy of Engineering.
Source: www.raeng.org.uk/policy/engineering-ethics/ethics#statement

In the publication 'Engineering Ethics in Practice' there are case studies that outline how these principles apply to real life situations. They illustrate how, for example, engineers need to get the right balance between the economic demands of the companies that they work for, and ethics. A recent example of this is the problems that Volkswagen have had with software that affected the readings for emissions from some of their diesel vehicles. Other ethical issues involve surveillance, cutting costs to ensure that projects finish on budget, or the use of poor quality or dangerous materials.

Reproduced courtesy of the Royal Academy of Engineering.
Source: www.raeng.org.uk/publications/other/engineering-ethics-in-practice-full

The environment

Almost everything that engineers produce, design or operate has a potential impact on the environment. Some large-scale projects have an obvious effect – a new dam, wind farms, new housing or office developments. Cars and aeroplanes produce emissions as a result of converting fuel into kinetic energy. Power stations produce electricity from fossil fuels or nuclear reactions and the by-products of these can be harmful to living things or have long-term effects on climate. Discarded batteries and electronic devices contain poisonous chemicals. Engineers are increasingly having to incorporate safeguards into their designs as public pressure is put on them to reduce environmental damage and avoid the depletion of natural resources and habitats.

Green buildings

It is estimated that between 30% and 40% of energy use in developed countries is associated with buildings, most of which is used in heating, cooling and electrical systems. The concept of a 'green building' is a relatively new one. By describing a building as being 'green', we mean that it is designed to be as energy efficient as possible. This can be achieved by:

- using building materials that do not require large amounts of energy to produce – plastics, concrete and metals require significant amounts of energy, whereas natural materials such as wood require much less
- insulating buildings effectively to avoid heat losses
- designing buildings with natural ventilation, such as using natural convection to produce cooling, rather than relying on air conditioning
- aligning buildings so that sunlight can be used either as a source of heating or of electricity in photovoltaic cells
- using 'thermal mass' to absorb heat and then release it naturally
- re-using water through recycling facilities.

Hybrid vehicles

Hybrid vehicles are vehicles – cars, trains, trams – that have two separate sources of energy. While the idea is not new (the moped is a hybrid vehicle as it is powered by a motor or engine and by the rider), car manufacturers are now devoting enormous resources to new ranges of hybrid electric vehicles. Hybrid petroleum electric vehicles are powered by internal combustion engines that run on petrol, diesel or gas. They also contain heavy-duty batteries that are charged as the engine is running, and the car can be powered electrically during parts of its journey. Other hybrid vehicles use the engine to compress air or hydraulic fluid, which can then be used to drive the wheels, and there are trams that are able to use a combination of diesel fuel and overhead electric cables. What all of these have in common is that they reduce the use of fossil fuels (and so reduce greenhouse gases and pollution) and utilise energy that might otherwise be wasted when the car is running but not moving.

Other examples of how engineers are adapting to the needs of the environment include:

- the development of new energy-efficient materials, such as composites that make aircraft bodies lighter, or insulators that prevent heat losses
- electrical conductors that have low electrical resistance in order to reduce heating when conducting electricity
- stronger, lighter materials for electrical devices
- ways of extracting minerals or fuels from below the Earth's surface that have lower environmental impact
- new ways of producing energy that do not rely on fossil fuels (see below).

Engineering and energy

Engineers use energy to create their products, and to run them. Engineers also create the structures and processes that produce usable energy. Engineers, therefore, are at the forefront of the search to develop new energy sources. As well as being finite sources of energy (in other words, they will one day run out, or, at the very least, become uneconomical to extract), fossil fuels cause environmental damage and are closely linked to climate change and global warming. Many fields of engineering have an interest in developing alternative or renewable energy sources. While there are no set definitions of these words, they generally refer to energy sources that do not involve fossil fuels and/or those whose consumption does not deplete the planet's natural resources or reserves.

Alternative energy sources that are currently used commercially are:

- geothermal energy
- hydroelectric power
- nuclear fission
- solar power (solar panels to produce heat and photovoltaic cells that create electricity from sunlight)
- tidal power
- biomass (using bacteria to produce gas through the digestion of organic materials, making fuels out of natural materials such as palm oil)
- wave power
- wind power.

An alternative energy source that is in development is nuclear fusion.

There are a number of ways of categorising energy sources (see Tables 1 and 2).

The engineers who are primarily involved in developing alternative energy sources are electrical engineers, electronic engineers, bio-engineers, mechanical engineers, geo-engineers and chemical engineers.

Table 1 Categorisation of energy sources based on origin

Energy sources derived from the sun	Energy sources not derived from the sun
Solar	Geothermal
Wind (heat from the sun creates areas of low and high pressure, causing wind)	Nuclear fusion
Wave (winds cause waves in the sea)	Nuclear fission
Fossil fuels (coal, oil, gas, etc. were once living things)	Tidal
Hydroelectric (water sources, such as rivers, were created by precipitation of rain, caused by the sun's heating)	
Biomass	

Table 2 Categorisation of energy sources based on effect on source

Renewable*	Non-renewable
Solar	Fossil fuels
Wind	Nuclear fission
Wave	Geothermal
Tidal	
Hydroelectric	
Biomass (since although the individual sources cannot be used again once the energy is extracted, the supply can be maintained by, for example, planting more oil palms)	
Nuclear fusion (considered renewable since the likely source would be sea water, which is effectively inexhaustible)	

* By 'renewable' we generally mean that the source of the energy is unaffected by the extraction of energy. For example, every time coal is burned there is less coal remaining on the Earth, whereas if a wind turbine creates electricity it does not affect the source (movement of air, which is caused by solar heating).

Engineering business case histories

Try to keep up to date with stories associated with engineering businesses and entrepreneurs. Most new engineering projects and developments are driven by commercial companies that want to sell their products or processes, and while some new devices are developed by university research departments, these are often funded by industry.

The list below aims to provide you with a starting point for your research, as there will undoubtedly be new names that will become more relevant by the time you read this book:

- Apple's iPad, iPod and iPhone
- the Android, iOS and Windows operating systems for mobile devices
- Google
- inventors, scientists and entrepreneurs, such as Eoin Musk or Tim Berners-Lee
- SpaceX or SpaceShipTwo
- multinational construction companies, such as Arup
- aeronautical engineering companies, such as BAE Systems
- UK tech companies (for example, those based in 'Silicon Fen', such as ARM)
- biotech companies, such as Gilead Sciences
- oil and petroleum companies that are developing alternative or renewable sources of energy, such as Statoil
- alternative energy schemes, such as the proposed tidal lagoon in Swansea Bay
- building tall structures, such as the Burj Khalifa in Dubai or the Shard in London
- issues surrounding the excavation and distribution of rare earth metals
- fuel cells that produce electricity from hydrogen
- how engineers deal with natural disasters, such as floods, earthquakes and tsunamis
- environmental issues, such as the safe processing and storage of used fuel rods from nuclear power stations, or polluting metals used in mobile phones or computers.

7| Non-standard applications

Perhaps you are studying for a mixture of examination qualifications, or you have had a gap in your education. You may have already started a degree course in another discipline and want to change direction. This chapter applies to students who may be applying to university as mature students, perhaps with qualifications other than A levels or the equivalent, to international students who are applying from outside the UK and to those with disabilities.

Whatever your situation, the first thing you should do is make contact with some universities (either by telephone or via the email addresses given on the university websites) to explain your situation and ask for advice.

Changing direction

Not everyone at the age of 17 has a clear idea of where their future lies. Often, university or career choices are made on a whim, on the advice of parents, teachers or peers; or simply through lack of experience of the chosen discipline. People change, and luckily universities understand this and are used to dealing with it. Making the wrong university choice should not lock you into a study path or career that is unsuitable, unviable or unpleasant. The case study below illustrates this.

Student experience

'I had always wanted to study engineering but towards the end of sixth form I began to enjoy my biology lessons more than my physics lessons and therefore made a last-minute decision to take up biology with the added practicality of forensics at university. Unfortunately, I did not enjoy it as much as I had hoped since it contained very little of my favourite and strongest subject, mathematics. At the end of my first year at university I decided to change from forensic biology back to my original choice of mechanical engineering, and it was very easy to do so. Every student is allocated a personal tutor to provide support and answer any queries we may have. I had a very brief meeting with my personal tutor, who also happened to be my course director. She advised me to do a bit of research about which course I wished to change to and, if I was completely happy, then the next step would be to collect a 'course transfer' form from the faculty office. The form then simply needed to be signed by my current course director, my new course director and myself. I emailed the course director for the school of engineering and arranged a meeting; he reviewed my UCAS application for forensics and helped me to decide which course would be best for me. We discussed my options and the course structure for all different branches of engineering. I felt very confident with my choice and the necessary documents were signed in the meeting and within a few days I received a confirmation email about the change.'

Aliya Foster, Kingston University, MEng Mechanical Engineering

Mature students

Mature students (defined as students who are over 21 at the start of their proposed courses) are usually:

- applying with appropriate qualifications, for example A levels, but have not been to university and are now applying after a gap of a few years, or
- applying for a second degree, having graduated in a different subject, or
- applying with no A levels or equivalent qualifications.

If you come under either the first or second of these categories, you apply using the same route as first-time applicants. However, you should contact the universities directly to discuss your situation with them, and to get their advice. The structure of your personal statement will need to be different from that produced by a student who is still at school or college. It should include:

- a brief summary of why you are applying now and what you have been doing since you completed school or college education
- your reasons for the change in direction
- an explanation of any gaps in your education or work history
- a discussion of any appropriate skills or experiences gained in your previous jobs or degree studies.

Most universities encourage mature students who want to apply for entry to degree courses, taking into account their work experience and commitment as part of the entry criteria. Mature students have often left school without the appropriate academic qualifications for university entrance in order to start jobs or careers, in which case there are many Access courses in colleges around the country that specifically prepare mature students for higher education. Others may have studied at degree level in another, non-related, discipline.

Universities often encourage mature students to apply because they:

- bring valuable real-life experience to the faculty
- can be more mature in their study methods than school-leavers
- have had more time to think about what they really want to study
- have a better understanding of the links between study and work.

The best way to find out about acceptable Access courses is to contact the universities directly and ask which ones they recognise or recommend. The engineering institutes (see Chapter 11 for a list of these) provide details of schemes that allow students who have studied on apprenticeship programmes to progress onto a Higher National Diploma or degree course.

If you are applying for a degree course as a mature student without using the Access course route, you should:

- use the UCAS application system as described earlier in this book
- select 'no' to the question 'are you applying through your school or college' when prompted on the online application
- fill in the section on employment as comprehensively as possible, ensuring that there are no periods of time since you left school that are unaccounted for
- ask someone suitable (your current employer or a previous teacher) to act as your UCAS referee
- ensure that he or she knows what the universities require from the referee (you can point them towards the UCAS website, which has a section on information for referees)
- send a more detailed CV directly to the universities once you have received your UCAS number (and quote this on all correspondence with the universities).

It is important to emphasise that the universities are keen to recruit serious, motivated and committed students onto engineering programmes, and mature students tend to fit this description extremely well. You will find that the university engineering departments are eager to help suitable candidates apply and they will be able to provide you with advice and feedback if you contact them prior to applying.

International students

In the 2016 application cycle, UCAS (www.ucas.com) applications and acceptance statistics showed that 2,200 EU students and 4,600 non-EU international students gained places on engineering courses.

International students are usually:

- following A level (or equivalent) programmes either in the UK or in their home country, or
- studying for local qualifications that are recognised as being equivalent to A levels, in their own country, or
- studying on academic programmes that are not equivalent to A levels.

Students following A level or equivalent programmes should apply through UCAS in the usual way. All of the information in this book is equally applicable to them.

Students studying qualifications that are accepted in place of A levels can also apply through UCAS in the normal way, from their own country. The UCAS website (www.ucas.com) contains information on the equivalence of non-UK qualifications. These include the Irish Leaving Certificate, European Baccalaureate, and some international O levels and A levels. Information on the equivalence of other qualifications can be found on the UK government's qualifications website (http://ecctis.co.uk/naric).

Students who do not have UK-recognised qualifications will need to follow a pre-university course before applying for the degree course. These include:

- university foundation courses that are offered by many UK colleges and universities. For example, the NCUK (www.ncuk.ac.uk) programme includes specialist science and engineering foundation paths. Typically, these programmes last nine months and involve 20–25 hours of intensive tuition a week, plus specialist English help
- university foundation courses set up by, or approved by, UK universities or colleges, but taught in the students' home countries
- A level courses (normally two years, but in some cases can be condensed into one year) in schools and colleges in the UK. A levels allow students to apply to all UK universities, including the top-ranked universities such as Oxford, Cambridge and Imperial College.

Foundation (or 'pathway') courses are not accepted by all UK universities. You should check with your preferred universities which courses they accept before committing yourself. Representatives of UK universities, schools and colleges regularly visit many countries around the world to promote their institutions and to give advice. You can also contact the British Council to get help with your application.

Non-EU students should note that they are required to pay higher fees than UK or EU students. UK and EU students have their fees capped at a maximum of £9,250 per year, whereas fees for international students on engineering courses are likely to be between £15,000 and £25,000 per year. Accommodation and meals will be extra. How much living costs are depends on where you study, but, as a rough guide, about £1,000 a month should cover food, accommodation, books and some entertainment costs.

Currently, EU students are still classed as 'Home' students and there will be no fee implications for students for 2017 entry; EU students starting their course in September 2017 will pay Home fees for the duration of their course. EU students starting their courses in September 2017 will remain eligible to apply for student funding under current terms. This was confirmed by Jo Johnson, Minister for Universities and Science, in September 2016.

Once Article 50 of the Lisbon Treaty is triggered, this will launch the process of formally negotiating Britain's exit from the EU. Therefore, for the time being, until it is announced otherwise, EU students will continue to pay the same fees as Home students.

One of the reasons why international students are, on the face of it, less successful in their applications than UK students is that they, their teachers or their referees are unaware of what is required, particularly if they have experience of applications for universities in other countries. All of the information on personal statements and interviews in this book applies equally to UK and international students.

If you are applying from outside the UK, you must ensure that your referee also understands what he or she needs to write. The UCAS website (www.ucas.com) contains a section giving advice to referees. In general, your referee should address the following points:

- some background information about you – a brief overview of your recent education, and why you are applying for a place at a UK university
- some background information about themselves – what their relationship is to you, how long they have known you, and on what basis are they able to comment on your academic and personal qualities
- an assessment of your academic and personal strengths
- an assessment of your suitability for a course and career in engineering.

English language requirements

Students whose first language is not English will be required to demonstrate that they have the right level of English language to cope with the course. Each university will specify a minimum score on an approved

SELT (Secure English Language Test) such as IELTS (International English Language Testing System). Students who have sat IB or IGCSE may not be required to sit an extra English language test. The UCAS Guide for International Students has details of English language requirements: www.ucas.com/ucas/undergraduate/getting-started/international-and-eu-stduents/tips-international-applications.

Checklist for international applicants

1. Check that your academic qualifications are accepted by universities. The UCAS Tariff lists all international qualifications that, in theory, are accepted. If your qualifications are not included in the Tariff, contact the universities directly.
2. Check that the level of your qualifications (grades, marks, predicted grades or predicted scores) are at the level required by the universities. The Course Search facility on the UCAS website will give you links to the universities' grade/score requirements.
3. Go to the international pages of the university websites to see if there are any specific language requirements, for example a minimum IELTS score. Some universities will specify an overall score only, for example 7.0. Others may also have particular requirements for each section of the test, such as 7.0 overall with at least 6.5 in each section.
4. Use the international pages to see whether there are any scholarships available to international students from your country.
5. Follow the advice in the engineering sections of the universities' websites regarding any specific information they want to see included in the personal statement.
6. Use the university websites to check the fee and accommodation arrangements.
7. If you are studying at an international school that has sent students to UK universities in previous years, it will be aware of its role in the UCAS application. The school is likely to already be registered with UCAS. If you are studying at a local school that has not sent students to UK universities, you will need to register as a private candidate on the UCAS website, and discuss the reference with someone suitably qualified to write it, perhaps one of your teachers or a previous employer.
8. Be clear about the deadlines for applications.
9. Check whether you need a visa to study in the UK (see below).

Visa requirements

Students from non-EU countries generally need a visa to study at degree level in the UK, although students from some countries are exempted. The UK government website has full details of whether you will need a visa (www.gov.uk/check-uk-visa). The stages in obtaining a visa are set out below.

1. You have been made an offer from an institution that is able to act as a sponsor for your visa application (www.gov.uk/government/publications/register-of-licensed-sponsors-students).
2. You have accepted the offer and have fulfilled the academic and English language requirements.
3. The institution has given you a Confirmation of Acceptance for Studies (CAS).
4. You apply for the visa, presenting the required evidence. Academic evidence such as an IELTS certificate or examination results will be listed on the CAS, and the gov.uk website will list other evidence, such as proof of funds to cover the tuition and living costs.

Note: Following the UK's decision to leave the EU, visa arrangements for EU students applying to UK institutions are to be negotiated as part of formal discussions with the EU, but for the time being no visa restrictions apply for EU students.

Students with disabilities and special educational needs

Universities welcome applications from students who have physical or other disabilities or special educational needs, and they have well-established support systems in place to provide assistance and special facilities. The services offered can help with a wide range of disabilities including sensory (visual/hearing) impairment, mental health difficulties, mobility impairment, dexterity impairment, Asperger's syndrome or other autism spectrum disorders, chronic medical conditions (e.g. diabetes, epilepsy or asthma) and specific learning difficulties (e.g. dyslexia or dyspraxia). In all cases, you should contact the universities directly, before you apply, to explain your particular needs and requirements. They will then be able to give you information on how they can help you. The website www.disabilityrightsuk.org contains useful links and information.

8 | Results day

You have done all of the hard work – your personal statement, the interview, the examinations – and you are now waiting for your results, the results that will determine whether you have achieved what you need to take your university place. This chapter explains what happens when you get your results and, if you have achieved grades or scores that are either better or worse than expected, what other options are available to you.

When the results are available

- A levels – third week of August
- IB – early in July
- Scottish Highers – first week of August

Ask your school or college for the exact date and time that they will issue you with the results.

Whichever of the exam systems you are sitting, you need to act quickly if you:

- have missed the grades or scores that you require to satisfy your firm offer
- are not holding any offers and wish to apply through Clearing (see below)
- wish to use the Adjustment system (see page 102).

If you have gained the grades that you need to satisfy your firm choice – congratulations, you have your place! The university will contact you with confirmation of the place.

What to do if things go wrong during the exams

Occasionally, students will underperform in an examination through no fault of their own. This could be through distressing family circumstances (a serious illness to a family member, for example), illness in the run-up to the exam (or during the exam) or unforeseen circumstances such as late arrival to the exam due to problems with public transport. In all cases, you should inform the universities that this has happened to you immediately after the examination. You should, if possible, get your referee to give the details to the universities and provide documentary evidence, such as a letter from your GP.

What to do if you have no offer

Students who are not holding any offers when the examination results are published (or those who have failed to achieve the grades that they need) can apply for vacancies through Clearing. The Clearing system operates by publishing all remaining university vacancies on the UCAS website and in the national newspapers. Students can then find appropriate courses and apply directly to the university. This time, you do not go through UCAS. UCAS will send you a Clearing Passport, which, when you have been made a verbal offer that you wish to accept, you then send to UCAS to confirm the place. Bear in mind that Clearing places at top universities are scarce, and so you will need to act very quickly.

What to do if you have an offer but miss the grades

If you have gained grades that nearly meet those required for your firm choice (e.g. BBB for an ABB offer) but are good enough for your insurance offer, your first choice can still accept you. Otherwise, you are automatically accepted onto the insurance place. Check on Track to see if you have been accepted. If not, contact the university and see if it can be persuaded to accept you – see sample email on page 102. Your referee might be able to help with this.

If your grades are well below those required for your firm choice but satisfy your insurance offer, you will be automatically accepted onto the insurance place. Check on Track to see if the insurance offer has been confirmed. If there seems to be a delay, contact the university.

If your grades are below those needed for the insurance offer, you are now eligible for Clearing (see above). Use the UCAS website and national newspapers to identify suitable courses from the published vacancies and contact the institutions by telephone.

If you have missing results (for example, an 'X' on your results slip rather than a grade) this probably means that there is an administrative error somewhere, for example a missing coursework mark. Contact your school or college examinations officer immediately to sort out the problem. Contact your firm and insurance choices and explain the situation to them and ask them to hold your place until the problem has been resolved.

To: anthony.john@allington.ac.uk
From: Jonathan Luke
Subject: A level results

Dear Dr John
UCAS no. 08–123456–7

I have just received my A level results, which were:
Mathematics A*, Physics A, Chemistry C.

I hold a conditional offer from Allington of ABB and I realise that my grades may not meet that offer as although I gained a higher grade in mathematics, I dropped a grade in chemistry. Nevertheless I am still determined to study engineering and I hope you will be able to still consider me for a place.

My Director of Studies is emailing you a reference. Should you wish to contact him, his details are: Mr S. Buckley, tel: 0123 456 7891, email: s.buckley@edinburghhs.sch.uk.

Yours sincerely

Jonathan Luke

If you have achieved grades that are not good enough to get you a Clearing place, you can be accepted onto Access, diploma or foundation places, and then progress to a degree course. Alternatively, you can resit your A levels and reapply next year. Discuss these options with your school and your parents. Don't make hasty decisions – ask the university to extend its deadline if necessary.

What to do if you meet your offer but you've changed your mind about the course

If your grades satisfy one of your offers but you have changed your mind about the course you want to study, you can be considered for Clearing courses if you withdraw from your firm/insurance place. Contact UCAS to withdraw from your original place. Use the UCAS website and national newspapers to identify suitable courses from the published vacancies and contact the institutions by telephone.

What to do if you exceed your expectations and offer requirements

Students who have achieved grades that are better than the predictions on which both their firm and insurance choices were based have the opportunity to enter Adjustment. If this happens to you and you decide you would like to apply to an institution that asks for higher grades, you can put your application on hold for a short time (a week) in order to see if any of these institutions would be willing to offer you a place. When you register for Adjustment you do not lose your original offer, so you are still able to accept this if you cannot find a course at another university.

Bear in mind that there are unlikely to be places available for Adjustment candidates on the most competitive courses, but it is still worth having a look if you are in this situation.

Retaking your A levels

If you have not achieved the grades that you needed for your chosen universities, and you do not want to take the available Clearing places, you could consider retaking one or more A levels. In the days when most examination boards offered January sittings, retaking might have meant studying for one term to boost the grade. The period from January to September could then be used to earn money, gain more work experience or travel the world. But apart from a very small number of exceptions, A level exams are now only available in June, and so retaking will involve studying for another year, so you need to be sure that your university aspirations are genuine enough to give you the motivation to add this extra year to your studies.

Speak to your teachers about the implications of retaking your exams. Some independent sixth-form colleges provide specialist advice and teaching for students. Interviews to discuss this are free and carry no obligation to enrol on a course, so it is worth taking the time to talk to their staff before you embark on A level retakes. Many further education colleges also offer retake courses, and some schools will allow students to return to resit subjects, either as external examination candidates or by repeating a year.

Reapplying

Universities are usually happy to consider students who are reapplying, either because they did not get the required grades first time around, or because they did not receive any offers of places. It is worth contacting the university to check whether this is the case. Some will have policies on grade requirements for retake candidates, while others might ask for evidence of any extenuating circumstances that may have affected the previous application.

Tips

- If there were extenuating circumstances that affected your application, include a brief mention of this in the personal statement ('I was disappointed not to have achieved the required grades, because my studies were affected by illness, but this has made me even more determined to become an engineer') but leave the details to the referee.
- If you are retaking, you can use the extra term or extra year to add weight to your application, for example by gaining more work experience, taking up a new subject, enrolling in evening classes that are relevant to your application, and furthering your reading.

9 | Fees and funding

The cost of studying at university comprises two elements: the tuition fees charged by the university and costs associated with accommodation, subsistence, travel, books, entertainment and other living expenses. The UCAS website (www.ucas.com) has full details of fees and support arrangements. This chapter explains how the tuition fees are calculated, and provides information on sources of funding for, or other help with, living costs.

Fees for undergraduate courses

The tuition fees that you will have to pay for undergraduate courses will depend on where you live and where you intend to study. From September 2017, universities will be allowed to charge home undergraduate fees up to £9,250 per year, as part of the government's new Teaching Excellence Framework (TEF), which will assess universities and colleges on the quality of their teaching. The higher-ranked universities will be able to charge the maximum amount, £9,250, though they are not unanimous in terms of when and whether they will effect this. Tuition fee loans will also increase to cover the higher fees. The fee cap for students studying in Wales remains at £9,000, while fees for students in Northern Ireland have yet to be confirmed for 2017 entry.

For UK students, the fees do not have to be paid at the start of each year. You are effectively given a loan by the government that you repay through your income tax once your earnings reach £21,000 a year for students in England and Wales, and £17,495 for students in Scotland and Northern Ireland. So if you never reach this threshold, you would not have to repay the fees.

- Students living in England are required to pay a maximum of £9,250 per year if they are studying in England, Scotland or Northern Ireland, and up to £9,000 if they are studying in Wales.
- Students from Scotland who study at Scottish universities are not required to pay tuition fees. They will have to pay fees of up to £9,250 if they study in England or Northern Ireland, and up to £9,000 if they study in Wales.
- Students living in Northern Ireland will pay up to £3,925 if they attend a university in Northern Ireland, up to £9,250 if they study in England or Scotland, and up to £9,000 if they study in Wales.

- Students from Wales pay up to £4,046 if they study in Wales, or up to £9,250 if they study in England, Scotland or Northern Ireland.
- EU students currently pay up to £9,250 per year to study at universities in England, £4,046 in Wales, and £3,925 if they study in Northern Ireland. Scottish universities do not currently charge tuition fees for EU students. Currently, EU students are still classed as 'Home' students and EU students starting their course in September 2017 will pay home fees for the duration of their course and remain eligible to apply for funding under the current terms. Following the UK's decision to leave the EU, fees for EU students studying at UK institutions beyond 2017 entry are to be negotiated as part of formal discussions with the EU.
- International students will pay higher fees, determined by each university (see page 97).

Support and information

England

There are maintenance loans of up to £11,002 per year available for students living in England. Further details can be found at www.gov.uk and www.hefce.ac.uk.

Scotland

Bursaries and loans of up to £7,625 a year are available from the Student Awards Agency for Scotland (www.saas.gov.uk).

Wales

Student Finance Wales (www.studentfinancewales.co.uk) provides loans up to £9,697 for students living in Wales. Learning and special support grants are also available.

Northern Ireland

Student Finance NI (www.studentfinanceni.co.uk) provides loans up to £6,780. Maintenance and special support grants are also available.

Postgraduate courses

Fees for postgraduate courses are determined by the individual universities, and will be different for home students and international students. For information about tuition fees, go to the university websites. Student loans for postgraduate courses are available from the Student Loans Company: www.slc.co.uk/services/postgraduate-loans.aspx.

Sponsorship

Engineering students are more fortunate than their peers who are studying other subjects, because of the large number of sponsorship and bursary schemes available from engineering institutes, companies and the universities themselves. This is because it is recognised that the UK needs to attract more able students into engineering. The starting points for finding out about sponsorship are:

- the university engineering departments
- the engineering institutes (contact details are given in Chapter 11).

The engineering institutes or institutions are professional bodies that accredit and represent their members, provide training and information, promote their particular fields of engineering, organise or offer scholarships and help engineers with their careers.

The level of sponsorship varies from course to course, university to university, and institute to institute, and also changes from year to year. You will need to spend some time researching your options. The benefits of gaining sponsorship are numerous. In addition to financial assistance, you will have opportunities to gain internships, work experience or even a job to go to once you graduate. The very fact that an organisation or university has been prepared to sponsor you also says a good deal about your personal or academic qualities and will enhance your CV.

Sponsorship can include:

- financial aid during the degree course
- paid or part-funded work placements during holidays or a gap year
- work or study placements overseas
- improved chances of jobs with the sponsoring companies after graduation.

Several publications giving details of the scholarships and bursaries offered by educational trusts are available, including the *Directory of Grant Making Trusts* by Gabriele Zagnojute. You should also refer to the Educational Grants Advisory Service (www.egas-online.org.uk).

Scholarships

There are a variety of scholarships available for engineering students – possibly more than for any other discipline. The professional engineering institutes also offer some scholarships – for example, the Institution of Mechanical Engineers, mentioned at the end of the chapter. You can also use the website www.scholarship-search.org.uk to look for undergraduate and postgraduate scholarships.

University scholarships broadly fall under three categories:

1. scholarships offered by the university
2. scholarships offered through donations by alumni
3. industry scholarships, which can include work placements.

Normally, students apply for scholarships once they have accepted the offer of a place from the university, although sometimes universities will offer a scholarship alongside the offer of a place in order to attract the best students. There is a section on scholarships under the 'Fees and funding' heading on all university websites.

The Institution of Mechanical Engineers

Undergraduate scholarships and grants offer assistance to students who are about to start or have already started on mechanical engineering degrees accredited by the Institution. Current undergraduate scholarships are sponsored by AMEC, Rolls-Royce Land Rover and The Eaton to assist students with their accredited degree-level programmes.

- IMechE Undergraduate Scholarships
- IMechE 'AMEC' Undergraduate Scholarship
- IMechE 'Land Rover Spen King Sustainability Award'
- IMechE Eaton Undergraduate Scholarship (£2,000 per year for up to four years)
- Whitworth Scholarship up to £20,000 over four years
- Overseas study award up to £2,000
- Group project award up to £500.

Source: www.imeche.org.

10 | Further training, qualifications and careers

There is no 'typical' day for a working engineer. One of the attractions of engineering as a career is that you are, with suitable planning, able to follow a career path that suits your own individual skills and ambitions. If you like working outdoors, working as an on-site civil engineer would allow you to do this, whereas if you enjoy working in a laboratory then you might choose to be a structural or electronic engineer. As we saw in the introduction to this book, engineers can work as part of a large team for a multinational engineering company or on their own in their own company. Engineers who are interested in planning and finance can work as product engineers, assessing the economic viability of manufacturing a product and then designing the production line.

Some students with engineering degrees decide to change direction after graduating. For example, engineers are highly sought after in the financial sector because an engineering degree demonstrates that the student has analytical and problem-solving skills.

The engineering institutes (see list in Chapter 11) and university engineering departments provide a good starting point for further investigation of possible careers through the case histories that they publish. Almost all of the engineering organisations have sections on their websites (often under the 'Education' tab) that contain profiles of undergraduate and qualified engineers. Some good examples can be found:

- on the Royal Academy of Engineering's website: www.raeng.org.uk/education/what-is-engineering/engineer-case-studies
- under 'Who are civil engineers' on the education pages of the Institution of Civil Engineers website (www.ice.org.uk/careers-and-professional-development/what-is-civil-engineering/who-are-civil-engineers)
- on the Try Engineering website (www.tryengineering.org)
- at www.whynotchemeng.com (for chemical engineering)
- on the Tomorrow's Engineers website (www.tomorrowsengineers.org.uk).

The websites of the university engineering departments are also useful sources of case histories, and they often have links that allow you to address questions to undergraduate engineers.

In its booklet *Educating Engineers for the 21st Century*, the Royal Academy of Engineering provides an overview of what qualities future engineers will need to possess:

> *No factor is more critical in underpinning the continuing health and vitality of any national economy than a strong supply of graduate engineers equipped with the understanding, attitudes and abilities necessary to apply their skills in business and other environments.*
>
> *Today, business environments increasingly require engineers who can design and deliver to customers not merely isolated products but complete solutions involving complex integrated systems. Increasingly they also demand the ability to work in globally dispersed teams across different time zones and cultures. The traditional disciplinary boundaries inherited from the 19th century are now being transgressed by new industries and disciplines, such as medical engineering and nanotechnology, which also involve the application of more recent engineering developments, most obviously the information and communication technologies. Meanwhile new products and services that would be impossible without the knowledge and skills of engineers – for instance the internet and mobile telephones – have become pervasive in our everyday life, especially for young people.*
>
> *Engineering businesses now seek engineers with abilities and attributes in two broad areas – technical understanding and enabling skills. The first of these comprises: a sound knowledge of disciplinary fundamentals; a strong grasp of mathematics; creativity and innovation; together with the ability to apply theory in practice. The second is the set of abilities that enable engineers to work effectively in a business environment: communication skills; team-working skills; and business awareness of the implications of engineering decisions and investments. It is this combination of understanding and skills that underpins the role that engineers now play in the business world, a role with three distinct, if interrelated, elements: that of the technical specialist imbued with expert knowledge; that of the integrator able to operate across boundaries in complex environments; and that of the change agent providing the creativity, innovation and leadership necessary to meet new challenges.*
>
> *Engineering today is characterised by both a rapidly increasing diversity of the demands made on engineers in their professional lives and the ubiquity of the products and services they provide. Yet there is a growing concern that in*

the UK the education system responsible for producing new generations of engineers is failing to keep pace with the inherent dynamism of this situation and indeed with the increasing need for engineers.

Reproduced courtesy of the Royal Academy of Engineering
Source: www.raeng.org.uk/news/publications/list/reports/Educating_Engineers_21st_Century.pdf

Chartered Engineer status

Chartered Engineer (CEng) is a professional title registered by the Engineering Council. Engineers who achieve this status have been able to demonstrate that they have reached a high level of professional competence. Attaining the status of Chartered Engineer brings many benefits, including:

- being part of an elite group of highly qualified engineers
- professional recognition of your qualifications and attainments
- higher earnings potential
- improved career prospects
- international recognition of your academic and professional qualifications
- access to continuing professional training.

As shown in Figure 3, the normal route towards gaining the qualification is:

1. an accredited bachelor's degree (BEng)
2. an undergraduate master's degree (MEng)
3. membership of one of the professional engineering institutes
4. experience of professional practice.

It usually takes between eight and 12 years from the start of an undergraduate degree to reach CEng status. For more details, contact the Engineering Council or one of the engineering institutes (see the list in Chapter 11).

Master's courses

Many undergraduate engineering courses are four years in length, and lead to a master's qualification (MEng) rather than a bachelor's degree (BEng). The alternative route to a postgraduate qualification is by taking a self-contained MSc course after completing the bachelor's degree. Master's degrees allow students to focus their studies on a specific area of engineering. Applications for self-contained postgraduate

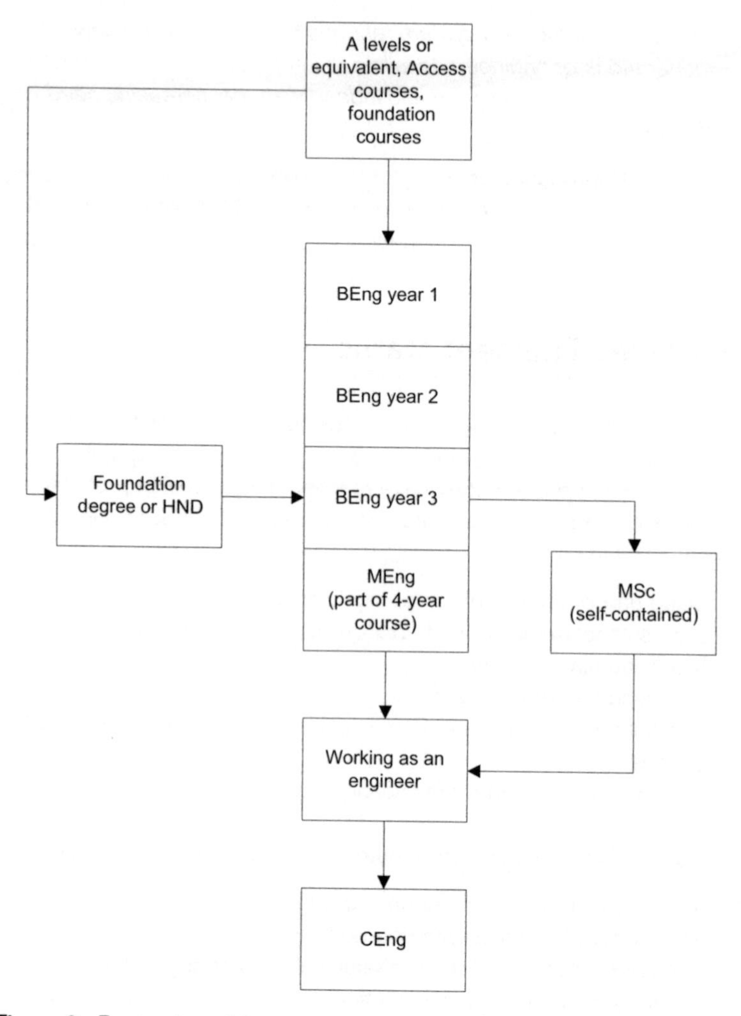

Figure 3: Routes to gaining an engineering qualification

courses are usually made directly to the universities, rather than through a central scheme. For listings of master's courses, see www.prospects.ac.uk and the university websites.

The advantages of doing a master's course are:

- greater specialisation
- better job prospects
- higher earning potential.

The cost of a master's course varies from university to university, but is likely to be in the range of £15,000–£23,000 per year for tuition fees. A number of grants and scholarships are available (see page 108).

The structure of a master's course will depend on where and what you study. As an example, here is the course programme for the MSc Energy Engineering with Environmental Management at the University of East Anglia:

- Electrical energy generation, distribution and storage
- Energy engineering dissertation
- Energy engineering fundamentals
- Oil and gas engineering
- Wind energy engineering
- Options: including energy futures; wave, tidal and hybrid energy engineering.

Source: www.uea.ac.uk.

Career opportunities and employment prospects

The engineering industry contributes around 25% of the UK's economic turnover, and employs nearly 6 million people in over half a million businesses. Employment prospects for UK engineering graduates are good, and as the UK economy recovers from recession, the manufacturing and engineering sectors look set to grow significantly. The organisation Engineering UK estimates that the UK will need over two million more engineers within the next five to 10 years.

According to figures produced by the Higher Education Careers Service Unit (HECSU) and the Association of Graduate Careers Advisory Services (AGCAS) for *What Do Graduates Do*, employment rates for recent engineering graduates were higher than the average for all graduating students, and the unemployment rates lower. For example, around 79% of graduating civil engineering graduates found employment within six months of graduating, with a further 13% embarking on further study or combining work and study. These figures compare with the averages for all graduates of 76% employment and 6% unemployment. The report also highlights the attractiveness of engineering graduates to other employers, with significant numbers finding employment within the managerial, IT and financial sectors. Source: www.hecsu.ac.uk.

In its report on employment, Engineering UK highlights the higher-than-average starting salaries for engineering graduates. With a mean starting salary of over £25,000, engineering was the fourth highest sector and around 16% higher than the average for all graduates. As a comparison, the average starting salary for the art and design sector was £16,000. Source: www.engineeringuk.com.

11 | Further information

Useful contacts

University applications

www.ucas.com

Funding

www.gov.uk
www.slc.co.uk
www.hefce.ac.uk (England)
www.saas.gov.uk (Scotland)
www.studentfinancewales.co.uk (Wales)
www.delni.gov.uk (Northern Ireland)

News

www.bbc.co.uk/news
www.theguardian.com

Overseas voluntary projects relevant to engineering

www.starfishgroupasia.com

Organisations for engineers in the UK

Engineering Council
www.engc.org.uk

Institution of Civil Engineers (ICE)
www.ice.org.uk

Institution of Chemical Engineers (IChemE)
www.icheme.org

Institution of Mechanical Engineers (IMechE)
www.imeche.org

Institute of Electrical and Electronics Engineers (IEEE)
www.ieee.org

Institution of Structural Engineers (IStructE)
www.istructe.org

Royal Academy of Engineering
www.raeng.org.uk

Chartered Institute for IT
www.bcs.org

Chartered Institution of Building Services Engineers (CIBSE)
www.cibse.org

Chartered Institution of Highways and Transportation (CIHT)
www.ciht.org.uk

Chartered Institute of Plumbing and Heating Engineering (CIPHE)
www.ciphe.org.uk

Chartered Institution of Water and Environmental Management (CIWEM)
www.ciwem.org.uk

Institute of Acoustics (IOA)
www.ioa.org.uk

Institute of Cast Metals Engineers (ICME)
www.icme.org.uk

Institute of Highway Engineers (IHE)
www.theihe.org

Institute of Marine Engineering, Science and Technology (IMarEST)
www.imarest.org

Institute of Measurement and Control (InstMC)
www.instmc.org.uk

Institute of Materials, Minerals and Mining (IoM3)
www.iom3.org

Institute of Physics (IOP)
www.iop.org

Institute of Physics and Engineering in Medicine (IPEM)
www.ipem.ac.uk

Institute of Water (IWO)
www.instituteofwater.org.uk

Institution of Agricultural Engineers (IAgrE)
www.iagre.org

Institution of Diesel and Gas Turbine Engineers (IDGTE)
www.idgte.org

Institution of Engineering Designers (IED)
www.ied.org.uk

Institution of Engineering and Technology (IET)
www.theiet.org

Institution of Fire Engineers (IFE)
www.ife.org.uk

Institution of Gas Engineers and Managers (IGEM)
www.igem.org.uk

Institution of Royal Engineers (InstRE)
www.instre.org

Nuclear Institute (NI)
www.nuclearinst.com

Royal Aeronautical Society (RAeS)
www.aerosociety.com

Royal Institution of Naval Architects (RINA)
www.rina.org.uk

Society of Environmental Engineers (SEE)
http://environmental.org.uk

Society of Operations Engineers (SOE)
www.soe.org.uk

The Welding Institute (TWI)
www.theweldinginstitute.com

Specialist publications

Architect Magazine
www.architectmagazine.com

The Engineer
www.theengineer.co.uk

Aviation Week
www.aviationweek.com

Nano
www.nanomagazine.co.uk

Race Car Engineering
www.racecar-engineering.com

Engineering News Record
www.enr.com

New Civil Engineer
www.nce.co.uk

E&T [Engineering and Technology] Magazine
http://eandt.theiet.org

Electronic Engineering Times
www.eetimes.com

Structures Magazine
www.structuresmag.org

Books

Engineering

- Blockley, David, *Bridges: The Science and Art of the World's Most Inspiring Structures*, OUP, 2010.
- Brain, Marshall, *The Engineering Book: From the Catapult to the Curiosity Rover*, Sterling, 2015.
- Brenner, Brian, ed., *Don't Throw This Away!: The Civil Engineering Life*, American Society of Civil Engineers, 2006.
- Browne, John, *Seven Elements That Have Changed The World: Iron, Carbon, Gold, Silver, Uranium, Titanium, Silicon*, Weidenfeld and Nicholson, 2013.
- Dupre, Judith, *Skyscrapers: A History of the World's Most Extraordinary Buildings*, Black Dog & Leventhal Publishers Inc., 2008.
- Dyson, James, *Against the Odds: An Autobiography*, Orion, 1997.
- Eberhart, Mark E., *Why Things Break: Understanding the World by the Way it Comes Apart*, Three Rivers Press, 2004.
- Fawcett, Bill, *It Looked Good on Paper: Bizarre Inventions, Design Disasters and Engineering Follies*, Harper Paperbacks, 2009.
- Gordon, J.E., *Structures: Or Why Things Don't Fall Down*, DaCapo Press, 2003.
- Gordon, J.E., *The New Science of Strong Materials: Or Why You Don't Fall Through The Floor*, Penguin, 1991.
- Hart-Davis, Adam, *Engineers*, Dorling Kindersley, 2012.
- Michell, Tony, *Samsung Electronics and the Struggle for Leadership of the Electronics Industry*, John Wiley & Sons, 2010.
- Monroe, Randall, *What If: Serious Scientific Answers to Absurd Hypothetical Questions*, John Murray, 2015.
- Munroe, Randall, *Thing Explainer: Complicated Stuff in Simple Words*, John Murray, 2015.
- Open University (author), *Engineering: the Nature of Problems*, Open University, 2016.

- Petroski, Henry, *Invention by Design: How Engineers Get From Thought to Thing*, Harvard University Press, 1998.
- Sammartino McPherson, Stephanie, *Tim Berners-Lee: Inventor of the World Wide Web*, Twenty-first Century Books, 2009.

Thinking skills

- Butterworth, John and Thwaites, Geoff, *Thinking Skills*, CUP, 2005.
- Tanna, Minesh, *Think You Can Think?*, Oxbridge Applications, 2011.

Glossary

UCAS applications

Adjustment
Runs in August and allows students who have performed better than expected to look at upgrading their university places.

Admissions tutor
Someone within an engineering department who deals with UCAS applications.

Clearing
The period from early July to mid-September when students who are not holding offers for undergraduate university places can approach universities that still have vacancies.

Deferred entry
Applications for an undergraduate place for the following year, allowing the student to take a gap year.

Extra
Allows students who are not holding any offers to approach extra universities, prior to receiving their examination results.

Gap year
A year between leaving school or college and starting university, usually used to gain further work or life experience or extra qualifications.

IELTS
The International English Language Testing System. Students who do not have English as their first language must reach a certain IELTS level in order to gain entry to study in the UK.

Legacy and new specification A levels
From September 2015, new A level specifications, in which the A level is examined at the end of the course rather than combining AS and A2 scores, were introduced and are being phased in over a four-year period. The full A level examinations for the first group of reformed subjects will be sat in June 2017.

UCAS
The Universities and Colleges Admissions Service, the online undergraduate application system.

Engineering

Aerospace engineering
A specialist branch of mechanical engineering focusing on aviation.

Automotive engineering
A specialist branch of mechanical engineering dealing with transport.

BEng
The qualification gained after a three-year undergraduate engineering degree course (often four years in Scotland).

Biomedical engineering
Linking engineering and living things.

Chemical engineering
The branch of engineering that looks at industrial processes involving chemicals, drugs, food and fuels.

Civil engineering
The branch of engineering dealing with large-scale infrastructure projects such as roads, bridges and dams.

Electrical engineering
Engineering involving electrical devices. Often taught as a joint degree with electronic engineering.

Electronic engineering
The branch of engineering that deals with electronics, such as integrated circuits.

Energy engineering
Engineering associated with energy production and transmission, environmental issues and alternative energy sources.

Mechanical engineering
The branch of engineering dealing with machinery.

MEng
The qualification gained from a four-year engineering degree course (often five years in Scotland).

MSc
A self-contained master's postgraduate degree.

Production engineering
Engineering that relates to manufacturing processes.

Structural engineering
Engineering that deals with the use of suitable materials for engineering projects.